THE OIL FINDERS

A COLLECTION OF STORIES ABOUT EXPLORATION

Collected and Edited

by

Allen G. Hatley, Jr.

CENTEX PRESS
Utopia, Texas

Copyright © 1995 Centex Press. Printed and bound in the United States of America. All rights reversed. No part of this book may be reproduced or transmitted in any form or by any means, electronic or mechanical, including photocopying, recording, or by an information storage and retrieval system–except by a reviewer who may quote brief passages in a review to be printed in a magazine or newspaper–without permissin in writing from the publisher. For information, please contact Centex Press, P.O. Box 510, Utopia, Texas 78884.

Although the Editor, Authors, and Publisher have made every effort to ensure the accuracy and completeness of information contained in this book, we assume no responsibility for errors, inaccuracies, omissions, or any inconsistency herein. Any slights of people, places, or organizations are unintentional.

ISBN: 0-9649416-0-0
LCCN: 95-083449

Photo on cover by Allen Hatley in 1976, of the testing of oil at Nido #1, offshore Palawan in the Philippines

Table of Contents

About the Editor iv

Foreword v

Introduction viii

About the Authors xiv

The Rover Boys and Other Stories 1
    By Dave R. Kingston

Ekofisk: A Giant Discovery of Oil and Gas in the North Sea 27
    By Ward W. Dunn

An Indonesian Experience 49
    By Donald F. Todd

Finding Oil Where It Shouldn't Be 119
    By Allen G. Hatley

The Story of Marinex and the Humbly Grove Discovery 141
    By John C. Kinard

Beyond Khartoum: Petroleum Exploration and Discovery 157
    By Allan V. Martini with James L. Payne

First Oil in the Sind 191
    By Herb Young

Miracles At Elmworth 218
    By John A. Masters

Yemen Oil Hunt—A Gift From the Gods 251
    By Ray Fairchild

About the Editor

Allen G. Hatley, Jr. has had a somewhat unique career, in that he sandwiched 10 years of working as a consultant and for independent oil companies, between working for two large major oil companies, initially Exxon and then Cities Service. This was accomplished during a period of time in the petroleum industry when such career changes seldom occurred. There followed another 10 years, again working for various independents and as a consultant, in addition to pursuing a second career outside the petroleum industry.

After serving in the U.S. Army, much of it in Korea and Japan, he graduated from Texas Tech University with a B.S. and M.S. in Geology. In 1955, he joined Humble Oil & Refining Company in Corpus Christi, Texas and later was assigned to the King Ranch Production District. After several years he was transferred to the Standard-Vacuum Oil Company, another Exxon subsidiary, and was stationed in the Philippines and in Somalia. In 1961, he joined Cabeen Exploration Corp., and for six years worked in the Philippines and in Peru.

While in Peru, he contracted Yellow Fever. He left Cabeen Exploration and spent almost two years consulting for several groups, including the World Ventures Group in Ecuador, prior to joining King Resources in 1969. In 1972, Allen Hatley joined Cities Service Oil Company, and his first assignment was to establish a one-man New Ventures Exploration office in Singapore. From 1974 to 1977, he was concurrently General Manager for Philippine Cities Service, Inc., where he is credited with recognizing the play type, acquiring the leases and making the first discovery of commercial oil or gas in the Philippines.

Upon assignment to Houston in 1978, Allen Hatley was appointed Manager of International Acquisitions for Cities Service—and then, following a major reorganization, Regional General Manager for East Asia, and finally Vice-President Eastern Hemisphere. He resigned in late 1982, after the Occidental Petroleum Co. merger was announced, and joined Gaffney, Cline & Associates, an international consulting company. In 1984, he headed The Western Company of North America's exploration subsidiary, Saturn Energy, during its most active period, and then was President, CEO, and a Director of Murexco Petroleum Co. in Dallas, Texas.

In 1987, he left Murexco and has spent the last 6 years working at a series of jobs, including a number of management and technical consulting assignments for various groups within the petroleum industry, as a Narcotics Agent in a federally funded Narcotics Task Force near the Mexican border, as an author of numerous articles and with one book in preparation, and as a Criminal Investigator for a District Attorney's office in central Texas. He holds a Texas Peace Officer's License.

Foreword

The idea for this book grew from a desire to document some of the significant petroleum exploration accomplishments from the 1960s through the 1980s. In 1988, I began to call a number of people who, I knew, could accurately tell the stories of the significant exploration efforts in which they and their company had been involved. The contributions in this collection are the result.

This collection of first-hand accounts was first published in a very limited edition in 1992. Here they are finally published in a new format for a wider audience. All are original accounts that relate, sometimes in an unusual chronicle, some great exploration adventures. They are also the ultimate tales about the understanding of petroleum geology and the busines of petroleum exploration.

In compiling this book, the authors of the individual stories and I, as editor, had complete freedom from censorship. None of the authors were told what to write or what not to write about. Each author was asked to tell his story in an entertaining and candid manner, and to tell as much about *why* he went looking for oil and natural gas in a certain area, as to tell *how* he did it. The authors were also asked not to make this a "geological textbook." As you read the stories, and, hopefully, live these adventures of exploration and discovery, I believe you will appreciate the unique way in which each storyteller accomplished this request.

Allen G. Hatley Jr.
Utopia, Texas, U.S.A.

Introduction

Allen G. Hatley Jr.

It was sometime late in 1953 when Rusty Bell interviewed me for a job with Humble Oil & Refining Company. During the interview he told me, "The most important job a geologist has with Humble is to find oil." With that statement, he had me hooked. I had already been interviewed by several companies, but none had said a thing about finding oil; they had spoken only about what a great training program or retirement plan they had. I have always suspected that the others did not talk about finding oil because they were either engineers or professional recruiters, but Rusty Bell was a geologist. Later, when I received an offer of a job from Humble and a couple of other companies, there was no question whom I would work for—it was the company that wanted me "to find oil," and knew how to talk about it.

This book is about finding oil. It's not a series of testimonials about one oil company or another, because companies do not find oil, although they sometimes provide help to those who do. But the reality is that only *people* find oil. This book is also not just about technological developments, or even about science. It's a series of first-hand accounts by geologists and "deal makers" talking about how they convinced themselves, their supervisors or investors, and various government officials that they could actually find oil—and about how they then went out and did it.

While working to collect these and other stories, some of which, unfortunately, you will not find in this book, I was amazed at the number of responsible people with important jobs in the petroleum industry who seemed afraid to let the public view real life in their companies. For example, as the stories in this book were collected, I talked to one geologist who, before he retired, had been deeply involved in an interesting major oil discovery in the Western Hemisphere. I thought the story belonged in this book, and he agreed to write the story.

Ultimately, the story was written, and when the author talked to his former supervisors (and friends) about publication, he was told that he would be sued if he published the story. I personally was shocked, as was the author, because I had read and heard a lot about this discovery and nothing he had written seemed either detrimental or proprietary in nature. Unfortunately, and much to my disappointment, neither that nor several other promised and very interesting histories found their way into this volume.

The authors whose stories do appear in this collection were mostly personal friends or long-term acquaintances of mine. From these connections, sometime in 1988 the idea was born to build a book from a collection of first-hand accounts of how a diverse group of geologists gave us some of the most significant discoveries of hydrocarbons in the last 25 years. But more than that, there was a strong desire to provide younger geologists reading these stories with a way to understand and learn something different about their chosen profession.

Some readers are bound to ask—Surely, there are some larger and more important oil and gas discoveries than those written about in this book? How did you select these—was it only because you knew the people who found the oil? Of course there were other significant and even larger discoveries made during the period covered, but knowing someone, or at least knowing who did what, and their knowing you, does help. Is this not the basis for that popular zip-word: "networking"? But there was much, much more to selecting real-life actors in the oil industry's greatest drama, than simply knowing their names.

First of all, the people who wrote these stories were articulate, had a sense of humor, and were not afraid to tell their story in a truthful and candid manner. Not a lot of people can pull that off, for a multitude of reasons.

Second, you will notice that many of these stories are rather easily traced to the accomplishments of a very few specific people. You will also notice that there are no testimonials to some fallen corporate chieftain or vast bureaucracy; this is because, in my opinion, a candid and true story will almost never appear in that type of story, as corporate politics corrupts all.

Last, the first-hand stories of the discovery of oil and gas related herein are unique and highly significant to the oil industry, the individual oil companies, and the nations touched by these oil finders' accomplishments. That is not just because some were very large discoveries; it is also because each story has something interesting to say that might be helpful to oil company supervisors as well as to many of the younger geologists trapped in today's largely non-risk-taking petroleum industry.

• **Dave Kingston** and the "Rover Boys" were among the last of an early breed of geologists and explorers who traveled to, and could confidently describe, the real "far away places" in the world, where you might want one day to explore for oil. Today, most geologists expect to be met at their foreign destinations by an air-conditioned car with driver. And few geologists today see any more of a new exploration area being evaluated than whatever is contained in the data room of the State oil company. Kingston, incidentally, after the initial field assignments he tells of here, went on to help develop and refine the world-famous Exxon Global Basin Classification System.

• **Ward Dunn** reminds us that even some major oil companies were once bold and proactive explorers who raced ahead of their competitors, many national governments, and even the signing of major international treaties, in their efforts to perform the first evaluations and acquire the first leases in unexplored basins. Ekofisk and other discoveries in the early days of North Sea exploration were the result of this drive to be "the first."

• **Don Todd** takes the reader into the chaos and uncertainty that existed in the Republic of Indonesia in the 1960s. He tell us for the first time how a very few Indonesian and American oil men not only negotiated the first of a new type of work contract that has become popular from Tunisia to Malaysia, but also organized a new exploration company, Independent Indonesian American Petroleum Company (IIAPCO). In 1966, while in Jakarta, Indonesia, Todd met Dr. Ibnu Sutowo, who was then the Mining Minister, Director General of Oil and Gas, and head of the State oil company, Permina. IIAPCO was then able to negotiate the first Production Sharing Contract. Eventually IIAPCO went on to survive political and operational problems and become the first of a new breed of highly successful independent international oil companies operating in Indonesia.

• **Allen Hatley** describes his company's inauspicious start in exploration in the Philippines in 1974. The story details the development of an imaginative, and for this offshore area, a new geological play concept; he then discusses some benign neglect by the home office, and finally describes the acquisition of new leases offshore Northwest Palawan. All this led to the drilling, in 1976, of a successful wildcat well only $3^1/_2$ miles from a previously drilled dry hole offshore Northwest Palawan. Fifteen years later, Northwest Palawan still contains the only commercial oil and gas fields discovered in the Philippines.

• **John Kinard** reminds us that sometimes bringing new exploration ideas to an "old exploration area" will pay off. With little or no experience in the area, Kinard and several associates undertook a geological study onshore southern England in 1971. Convinced there were undrilled prospects in the area, Kinard searched for backing to acquire leases and drill wells in an area where he was reminded, time after time, that the place had been relinquished by British Petroleum, "who were the best exploration company in the U.K." Finally, in 1976, his company produced oil at Humbly Grove, and this discovery acted as the stimulus for additional drilling and discoveries in southern England for the next decade.

• **Allan Martini** and **Jim Payne** tell about Chevron's long and eventful exploration program in one of the least understood, least explored, and most remote areas of Africa. In exploring an area only slightly smaller than the state of Texas, Chevron spent 4 years conducting a major exploration effort both in the Red Sea and in the interior areas of the Sudan. In the interior, and after drilling four dry holes, Chevron drilled their first discovery, at Unity #2, in 1979. Other discoveries followed, and Chevron then began to work and develop ways to bring its newly found oil to market. Twelve years later, no hydrocarbons have been produced from these discoveries. The authors, who worked on different phases and in different areas on this huge exploration project, tell us about the problems, the triumphs, and the frustrations involved.

•**Herb Young** takes us from Karachi to the Khyber Pass, and back to the southern Sind Province in Pakistan, where in 1981 his company made the first discovery of oil in the region. Since that first small discovery, 36 additional oil and gas fields have been discovered in the same area. Young talks about his long involvement in Pakistan exploration projects, while he worked for several oil companies. He also shares with us how the idea for the geological plays originated, tells about the negotiations for the leases, and describes how numerous exploration problems were eventually solved.

•**John Masters** talks about what he has aptly called "miracles." His is the story of the discovery of the Elmworth gas field in Western Canada. This huge gas field was found in an area abandoned by virtually every large oil company in Canada, because the conventional wisdom at the time considered the area to be unfavorable for hydrocarbon entrapment. His is the story of exploration using geological concepts that were unrecognized by the petroleum industry at the time. To many, the discovery of huge gas deposits at Elmworth was truly a "miracle," because most of the so-called experts in the petroleum industry in Canada called Masters crazy and even discounted successful production tests. His is a story that again proves that oil and gas are truly found first in the minds of innovative "oil finders."

•**Ray Fairchild** relates much more than just the story of one of the major oil discoveries in the 1980s—he also gives us some insight into the beginnings of Ray Hunt's successful international exploration program. We are reminded that successful exploration starts with the ability and desire to take big risks in frontier exploration areas, but even then it will only succeed if you have the right people—people who are not only experienced, but who also don't let a little adversity stop them when traveling to the out-of-the-way parts of the world.

All of these discoveries of oil and natural gas changed the lives of those who controlled the land on which the wells were drilled, just as they changed the lives of those who drilled the wells and made the discoveries. A number of companies have been born as the result of these discoveries, while existing companies and numerous governments have been impacted by these oil finders' efforts.

The nations of Pakistan, the Philippines, Indonesia, Yemen, Norway and, hopefully, one day the Sudan, owe much of their timely exploration activity of the 1970s and early 1980s to these discoveries. Can anyone considering exploring today in the rift basins of Africa believe they would be there had it not been for Chevron's discoveries in the Sudan? And does anyone really believe that exploration would be continuing in the Philippines had there not been a discovery at Nido?

For younger geologists and explorationists (my favorite name for a potential oil finder), I believe there is much to be learned from examining both the stories told and the attitudes and actions described by the "oil finders" writing in this book. For example, few of these oil finders were satisfied with just being good geologists. None were what today are called "specialists," practicing the newly found disciplines of earth science we see so highly touted in many of today's universities. Most of the oil finders in these pages were actually a combination of landman, negotiator, historian, scientist, teacher, and sometime political scientist. All were individually persistent, imaginative, daring, and occasionally abrasive. But above all, they were innovative salesmen of their ideas and dreams, or they would not be telling their stories in this or any other book.

If the primary measure of a petroleum geologist's success is his ability to explore for and discover new deposits of oil and natural gas, then the men telling these stories stand out among their peers as examples of the real "success stories" in their chosen profession.

Most of the oil finders in this volume traveled far from home to drill their wells. Many also spent considerable time in, or getting to and from, the more inhospitable regions of the world, whether in the offshore areas of the North Sea, Indonesia or the Philippines, or in the jungles, swamps and deserts of the Sudan, Pakistan, Yemen, Turkey, or interior Africa.

None really attained fame or fortune as younger geologists; all had served their time "in the trenches" and had worked more than a dozen years when they made their mark as an oil finder. Most of the oil hunters in this book have also avoided the large oil company Board Room—or maybe those on The Board avoided them. The exceptions, of course, are those who are presidents or directors of their own companies.

Lastly, it is important to understand that, at the time of their greatest exploration accomplishment, these oil finders often put their willingness to take a major professional and exploration risk ahead of personal comfort and safety, family, corporate politics, and sometimes monetary gain.

If all this sounds too romantic and out of character for the oil industry in the 1990s, I suggest you are correct. The authors in this book all lived in another place and time, even though some are relating stories of less than a decade ago. Nevertheless, they lived and worked in a "different world."

Theirs was a world where replacement of reserves was critical to success; it was a world in which exploration risk-taking was a natural part of the growth of most oil companies, and it was a world where geologists working for oil companies were expected to measure success by giving an affirmative answer to the question— "How much oil have you found, today?"

ABOUT THE AUTHORS

WARD W. DUNN

Ward W. Dunn, a Petroleum Consultant, resides in Adamsville, Rhode Island, having retired from Phillips Petroleum Company as Vice President Exploration in 1984. He graduated from Princeton University with a B.A. in Geology in 1942, and spent the following three years in the Pacific Theater in World War II attached to an aerial photography squadron as a photo interpretation officer in the U.S.N.R.

After being discharged from active duty as a Lieutenant in the U.S.N.R., Ward spent the following three years with the Gulf Oil Corporation in west Texas as a development geologist. He joined Phillips Petroleum Company as a subsurface geologist in 1948, and spent the next 11 years in the Rocky Mountains contributing to the discoveries of deep pay zones at Golden Eagle oil field and the discoveries of South Cole Creek and East Poplar oil fields in Wyoming and Montana.

In 1958, Ward joined the International Department of Phillips as Assistant Director Exploration and Production International Department. He subsequently was made Manager of the Paris office in 1962 and was instrumental in the discovery and development of the Hewitt gas field offshore Great Britain and Ekofisk offshore Norway. In 1973 he was transferred back to Bartlesville headquarters and became Vice President of Latin America and Asia Exploration and Production, where he directed development of fields in Indonesia, and offshore Iran, and exploration in Latin America, the Middle East, and the Far East.

In 1976, Ward Dunn was awarded the Royal Order of St. Olaf in the Degree of Commander by the King of Norway in recognition of contributions to the discovery and development of Ekofisk in the North Sea. In 1980 he was elected to Vice President Exploration for Phillips Petroleum Company. Ward Dunn passed away in 1994.

RAYMOND E. FAIRCHILD

Nineteen twenty three was a typically inauspicious year in which to be born, but Ray Fairchild shared the year of his birth with such events as President Harding's death, Mussolini's first year in power, and Pancho Villa's murder. Out of this melange of human events, and seasoned by elementary and high schools in Ohio, and then by Ohio University, the Johns Hopkins University, and the University of Missouri, Fairchild attained the requisite standard of education with only the minor interruption of World War II in Europe.

Finding employment in 1950 was at least as challenging as finding oil, but, endowed with an army insurance dividend and the sale of his Argus C-3 camera, Ray found work as a geological scout in Shreveport with Pan American Production Company. Later he worked in the Abilene and Houston Texas offices. In 1957, he joined the new exploration department at Trunkline Gas Company in Houston; later he became Gulf Coast Division geologist for Anadarko Production. In 1972 he was fired in the great "shoot-out at the Anadarko Corral," which in retrospect was probably one of his greatest career events. It lead to his entrance into international work at Maersk Oil and Gas in Copenhagen, where for $6^1/_2$ years he served as exploration manager for Maersk's onshore and offshore Danish territory, and for their international expansions into Germany, Colombia, and other parts of the world. Fairchild returned to the states at the end of 1979 and joined Hunt Oil Company in Dallas as international exploration manager, charged with successfully involving the company in international exploration efforts. The Marib basin play in North Yemen was the first of several he undertook before he retired as Senior Vice President, in June 1988.

Amongst professionals, one's career often becomes the standard of achievement and recognition, but in Ray's case, a somewhat different drummer sounded. His family has been and shall be "numero uno." His wife, Eleanor Faith, is well known in the international community. She joined Ray in marriage in 1973 in Copenhagen; their six children and five grandchildren share Ray and Eleanor's affectionate attention to their farm in Wood County, Texas.

ALLEN G. HATLEY, JR.

Allen Hatley graduated from Texas Tech University with B.S. and M.S. degrees in Geology. His experience consists of over 30 years in the international and domestic petroleum and minerals industry, 17 years of which were spent living outside of North America. Highlights of this experience are: He has worked for two integrated and several independent oil and gas companies as a paleontologist, a well-site, surface, and subsurface geologist, an International Land and Negotiations Manager, an Exploration Manager, an Operations Manager, a Regional General Manager, an Eastern Hemisphere Vice President, and a President and Chief Executive Officer for two independent oil companies. He also worked for 12 of those years as a consultant to numerous independent oil companies and for national governments.

During that period, Allen Hatley identified the geological play type, led the acquisition of exploration rights, and managed the operation that resulted in the first discovery of oil in the Philippines. Allen also managed the exploration program and led the acquisition of exploration rights for the group that identified the play type and mapped the structures that eventually became the Jabiru and other discoveries on the Northwest Shelf of Australia.

While Vice President of the Eastern Hemisphere for Cities Service Company, he managed all lease acquisition, exploration, and production in the Eastern Hemisphere, including the supervision of operations in 13 countries, and annual budgets greater than $120 million.

More recently, as a Partner in Energy Ventures Group, Allen led the efforts of several independent oil companies in a successful acquisition of a million-acre new exploration block in Tunisia, where he also will be involved in the exploration operations. He is currently a consultant to the World Bank and an Advisor to the Management of a large international consulting firm.

JOHN C. KINARD

John Kinard got his first introduction to the oil business while he was a teenager traveling with his landman father. John also worked for Core Lab and Texaco during his summers in college. He graduated from Oklahoma University with a B.S. and an M.S. in Geology, and in 1956 he joined Phillips Petroleum Company in Billings, Montana, as an exploration geologist. He left Phillips in 1960 and continued his career in the oil industry as a consultant and independent operator. In 1967 he formed Remuda Corporation, a company he still owns, and in 1970/1971 he helped form Marinex companies.

The Marinex companies have been active in several exploration ventures in various parts of the world, including the North Sea, Spain, and onshore United Kingdom. In 1980, while John was a Director and one of the owners of London-based Marinex Petroleum plc., the company acquired leases onshore southern England. It was on those leases that the Humbly Grove oil field was discovered.

John Kinard resigned from Marinex in 1983 and returned to the United States to manage Remuda Corporation. In association with others, he also formed a U.K. company that is now active in exploring in the North Sea and onshore southern England.

DAVE R. KINGSTON

Born in 1927, Dave R. Kingston was raised in the backwoods of Wisconsin. Cutting timber, trapping, hunting, fishing, and the usual high school sports occupied his early years. After military service he attended the University of Wisconsin, receiving his B.S., M.S., and Ph.D. degrees in Geology and Meteorology. During university summer vacations he explored northern Canadian mountain wilderness areas, describing and mapping the geology.

Dave joined Esso (now Exxon) in 1953, and spent the next 33 years working for that group, exploring the unknown oil basins of the world. The field parties sometimes worked in very primitive areas—out in the jungle or deserts, living off the land, separated from headquarters for as long as a year. These roving field parties led by Kingston were known as the "Esso Rover Boys." Tales of their exploits and experiences became legend both within the company and in the oil industry. For many years, the Rover Boys were responsible for the initiation of several new Exxon oil ventures that added 10 to 15 billion barrels of new oil to company reserves.

As a team, Kingston and his colleagues also developed the "Global Basin Classification System," which described all sedimentary basin types in the world and their possible petroleum potential. Their paper on the subject was published in 1983 and won the AAPG best paper award for that year. This system is now used by industry and foreign governments in worldwide petroleum appraisal.

Dave retired from Exxon in 1986, and is now an international petroleum consultant and teaches exploration schools for oil companies, foreign governments, and academic institutions.

ALLAN V. MARTINI

Born in Minneapolis, Minnesota, January 11, 1928, Allan V. Martini attended Minneapolis public schools. Following military service, he attended the University of Minnesota, graduating in 1951 with a degree in Geological Engineering. He began work as an exploration geologist with The California Co. (a Chevron forerunner) in Casper, Wyoming, in 1951, and worked for 17 years in various exploration assignments in the Rocky Mountains/Williston Basin, Louisiana Gulf Coast, and California.

In 1968, Allan became Vice President-Exploration of Western Operations Inc. (another Chevron forerunner). In 1973 he became Vice President-Exploration, and in 1980 he became President of Chevron Overseas Petroleum Inc., responsible for Chevron's foreign exploration/production operations in areas other than Canada, Indonesia, and Saudi Arabia. From 1984 through June 1986, Allan was Senior Vice President Chevron USA. In July 1986 he was elected a Director of Chevron Corporation, and as such was responsible for worldwide exploration/production activities. He retired from Chevron in August 1988.

Allan Martini married in 1951 and he and his wife, Eleanor, have four children (no geologists). Now at home in Tiburon, California, near San Francisco, Allan spends lots of time flying, both sailplanes and power.

JOHN A. MASTERS

John Masters is an exploration geologist. He was born 64 years ago in Iowa but grew up in Tulsa, Oklahoma. His father died when John was young, but he was blessed with an angel for a mother. He went to Yale on a scholarship, where he was educated as a gentleman in liberal arts; then he entered the cruel world by taking an M.Sc. in Geology at Colorado University.

John's first major discovery was at Ambrosia Lake, the largest uranium deposit in the United States. John became, at 32, the Chief Geologist for Kerr-McGee, and discovered Dineh-bi-Keyah, the largest oil field in Arizona, and helped find two new fields in Ship Shoal, Gulf of Mexico.

Then, John emigrated to Canada in 1967, as President of Kerr-McGee of Canada, and he finally terminated in 1973 when he declined to transfer back to Oklahoma City. He left his great mentor, Dean McGee, but started a new career with his partner and trusted friend, Jim Gray. They formed Canadian Hunter Exploration, got Noranda to fund them, and three years later discovered Elmworth, the largest gas field in Canada. Hunter, a technical leader in the industry, was built on that field.

The Masters family is Lenora and five children. All are expert deep-snow skiers and ski together in British Columbia each year. John taught all the children to climb, and he joins still in hiking, camping, sailing, windsurfing, and swimming, to name a few of his activities.

John reads widely every night from his personal library and still directs exploration for Hunter.

JAMES L. PAYNE

James L. Payne is Chairman, Chief Executive Officer, and President of Santa Fe Energy Resources, Inc., a large independent oil and gas company headquartered in Houston, Texas.

Jim graduated from Englewood, Colorado, high school in 1955, and received a Geophysical Engineering degree from the Colorado School of Mines in 1959 and an M.B.A. from Golden Gate University in 1974. He later attended the Stanford University Executive Program and is a Certified Professional Geologist.

Jim's first 23 years in the oil business were with Chevron, for whom he held a variety of operations and exploration management assignments for domestic U.S. regions, Europe, the Middle East, and Africa. Payne came to Santa Fe Energy in 1982 as Senior Vice President-Exploration and Land. He was named President of Santa Fe Energy Company in 1986 and was appointed Chairman and CEO of Santa Fe Energy Resources, Inc., in June 1990.

Jim Payne is active in the Society of Exploration Geophysicists, Spindletop International, Domestic Petroleum Council, and AAPG.

DONALD F. TODD

Don Todd was raised in Michigan, the son of a small-town businessman who spent (and lost) all his extra dollars seeking his fortune while looking for that big oil field. This was an era of cable tools and witch-stick doodle buggers, and he became fascinated with the elusive hunt for oil.

After World War II, Don attended the University of Michigan, graduating in 1950 with a B.S. degree in Geology. As no jobs were available due to an industry recession, he hung out his shingle in southern Kentucky, hoping to find oil at 200 feet.

Early in the Williston Basin boom, Todd joined Stanolind Oil & Gas in Billings, Montana. Destined to be independent, he resigned 18 months later, remaining in Billings as an independent geologist until opening IIAPCO's Jakarta office in 1967.

After IIAPCO's merger with Natomas Company in 1968, Don again became a public company employee. With feet itchy to be an independent once more, he resigned in 1971 to form Southern Cross Limited, living in Denver.

Todd was associated with Larry Barker from 1953 to 1978 in Tyler Oil Co., IIAPCO, Southern Cross, and Draco Mines. Tyler Oil had modest success in central Montana; IIAPCO and Southern Cross had many outstanding discoveries in Indonesia. Todd's career has included field work, well-site work, prospect generation (surface, photo, subsurface, and seismic), negotiations, selling, and management.

Montana was a good place to dream and to develop a frontier background. In the boonies, leases were available and cheap. There was little, if any, competition. Although obtaining risk capital for far-out exploration is always difficult, Don learned that the reward to risk for a frontier success was far greater than close in wildcatting. He gained confidence in himself and his belief that carefully thought out and logical frontier exploration wasn't that much more risky than normal wildcatting. That Montana philosophy worked in Indonesia and in all areas he worked after returning from Indonesia.

Don Todd's current semi-retirement involvement in the industry is through his companies, Constellation Group Limited and Toucan Oil Trust.

HERB YOUNG

Herb Young is a geologist who for 37 years has been engaged in petroleum exploration throughout the world. During his career, he has lived in seven countries in South America, Europe, South Asia, and the Far East. He recently retired as Managing Director of Union Texas Petroleum Limited in London. In that position, he had general management responsibilities in the United Kingdom and exploration responsibilities for Europe, Africa, and the Middle East.

Herb was born in 1929 in Great Neck, New York, where he lived until entering university. He enrolled in the School of Mines of the University of Idaho, with the intention of becoming a mining engineer. However, his fascination with prospecting led him to a degree in geology, which he attained in 1951.

After military service, an opportunity to work as a surface geologist in the Upper Amazon of Peru brought Herb into petroleum exploration. He has had many exciting adventures in his overseas career, with the most notable being his receiving a decoration from the Pakistan government in recognition of his services in the field of petroleum exploration.

Herb's military service as an officer in the U.S. Marine Corps included duty in Korea during the conflict and as Office-in-Charge of the Office of Ground Water Resources, Camp Pendleton, California.

Herb Young is married and has three children. He and his wife, Susan, met in an oil field camp in eastern Venezuela, where her father was Superintendent of Production. Susan was born in Wales but grew up in oil field camps in Trinidad and Venezuela—which prepared her well for a life of following the oil booms around the world.

Frontier Exploration

The Rover Boys and Other Stories

By Dave R. Kingston

They called us the "Rover Boys." That was short for the job title "Roving Geological Assignment, Worldwide," for the Standard Oil Company of New Jersey (once Esso and now Exxon). The Rover Boys were an elite group of geologists carefully selected and trained, who, during the 1940s through the 1960s, were sent out by Standard Oil (Esso) management into the unknown basin areas in the world to make quick assessments of oil prospects.

Never numbering more than three or four at one time, this group of geologists was able to operate in any kind of terrain: deserts, mountains, or jungles. Also required were unmarried status, a strong survival instinct, an ability to live off the land and withstand hardships, the ability to learn new languages, and a knack for getting along with the natives. We were supposed to be tough—"So tough," our management declared, "that [we] chew tobacco and spit into the wind."

Over a period of some 20 years, Esso Roving Teams high-graded the geologically unknown areas of the world, selecting the most promising for active exploration and down-grading the rest. After World War II, reconnaissance surface geology was the main tool used to predict exploration potential of new areas. That was where the Rover Boys came in. We covered the basin areas to locate the size, shape, tectonic features, and structures, and then worked out the stratigraphy, projecting surface rocks into the basin depths.

The Rover Boys were an institution within the Standard Oil Company of New Jersey. The first groups were formed in the 1940s when Esso headquarters decided they needed their own exploration group, separate from the affiliates. They sent out field parties under the late Maurice (Abner) Wallace to locate or check out new and promising areas. Maurice Wallace was called Lil' Abner after the comic strip character, since he was 2 meters (6'6") tall and weighed 114 kilos (250 pounds). His group worked in South and Central America and Africa. In the early 1950s, my team of Rover Boys took over—Shelby Eddington, Dick Murphy, Pat Shannon, and me. We roamed the globe, working on all continents, living out of duffel bags, always on the move. In

The Rover Boys in Africa, 1959, near Timbuctu, Mali Republic. From left: Dick Murphy, Dave Kingston, George Voutopoulos (our mechanic), and Shelby Eddington.

the 1960s we concentrated our studies by doing all the basins of Europe as a group. A little later we did the same for Africa.

By the mid-1960s, the best-looking onshore basins worldwide had been examined and surface field work had given way to a new frontier—the offshore. There, reconnaissance geology was done by seismic. The roving days were over by the end of the 1960s. Today, the surviving Rover Boys are unanimous in one respect—we think we had the best job in Esso, and perhaps in the whole oil industry. We had an opportunity to see the world, to do the job we loved most, and to do it with all expenses paid.

For many years, however, during the 1940s, 1950s, and 1960s, the Rover Boys provided the staff at Esso headquarters with tales of adventure and misadventure, some hilarious and some serious, in addition to a constant stream of new oil prospects. Headquarters shared these experiences with other company exploration personnel via their *Exploration Newsletter*, a monthly publication whose motto was "All the news that's fit to print, and some that's not."

While we were out in the field, reports on our operations and activities were sent back to headquarters in New York. These were labeled as monthly reports. Many of these reports were included in the

Newsletter, and some became quite famous within the company. The shortest was from Abner Wallace, who, having wasted a whole month getting customs clearance for his equipment in a remote African port, fired off the following as his monthly report to New York.

> To: Headquarters ESSO, NYC
> From: Abner Wallace
> Every day a holiday, every meal a banquet, every night a party.
> [signed] Party Chief

This report was received with much hilarity in New York and was hailed by management as the ideal for content and brevity. Private letters could also find their way into the *Newsletter*. We wrote the following to Bill Wallis, our exploration manager (1959), from our field party in the North African desert.

> Dear Bill,
> We hear you are going to fly to Paris in April, and have a request to make. Please bring along a couple of cans of WALNUT pipe tobacco, and we will pick them up when we arrive there in May. As you know, we have been out in the Sahara Desert since last fall. We ran out of regular pipe tobacco a couple of months ago. Since that time we have been smoking camel dung (dried, of course). Please understand that we are not complaining about having to smoke camel dung, but the fact is that we are beginning to enjoy it. Since our "desert tobacco" will be hard to find in Paris, we think we should get back to the regular stuff.
> (signed) D. R. Kingston

This, of course, went into the *Newsletter* verbatim, since all communications from the field were considered fair game.

Being a Rover Boy was a great job for young geologists. We thought it was the best in the world. We were exploring the unknown, doing geology, and camping out. We were seeing the world at company expense and we were getting paid as well. We lived out of duffel bags, and sometimes didn't eat regularly, but there were no complaints. For us it seemed almost too good to be true. We had complete freedom to do our jobs as we saw fit. Our basin studies took us to South America, the Middle East, Africa, and Europe. We did a fast mapping job on the basins, predicted the type and size of prospects, and then went on to the next job. We were told to go after the elephants. "Look for billion barrel structures." Gas was out—except in Europe or North America. "You lose your ass, finding gas," the saying went, because in those days there was no market for gas. Our job was to locate the "oily" basins and big structures, ahead of competition. Because we had to move fast, we couldn't spend too much time in one place. Our costs were low. Our New York management figured that the Rover Boys were the cheapest and most effective exploration tool that they had.

Other companies had field parties, too, but they were larger and more cumbersome. We followed another company who had gone into the Central Sahara in North Africa with an expedition of 20 vehicles, 50 men, and an airplane. It took them 2 years to get the job done. We went in with four men, two jeeps, and one truck, and finished in 5 months.

For transportation we used jeeps or trucks where there were roads or desert; horses, mules, camels, and even human porters carried our gear where no roads existed. In the tropics we also traveled by dugout canoe. Whenever possible, we would hire small airplanes to do aerial reconnaissance. Our objective was always to see as large an area as possible. Often, we were out in the desert or bush for long periods, sleeping on the ground, living on fish and game.

We did all of our own camp chores, cooking, and dishwashing, and there were no extra hands. The team consisted of two to four geologists and a mechanic if we were using jeeps. In the desert, the mechanic was the most important man—you could get along without one or two of the geologists, but if your jeep broke down and you couldn't fix it, you were dead. We took one of the natives along as interpreter if none of us could speak the local language, or if there were bandits and he knew who they were.

Very seldom did any of us get sick. We had our medical kits containing the "wonder drugs" developed during World War II; infections like malaria, yellow fever, and a host of other maladies of the past had become preventable. But the main trick in avoiding sickness on these long sojourns was to stay away from people, which was why we had no cooks or dishwashers with us and why we would always camp outside of villages, not in them. We avoided small town hotels and restaurants whenever possible. Civilization—cities, villages, and native encampments—were where water was bad, mosquitoes carried malaria, and where sickness was indigenous. Stay out in the bush and avoid people—that was our motto. In that way we avoided the bugs and most of the bandits.

Sometimes we were out of contact with civilization for 6 to 10 months. Our headquarters didn't know where we were on a day-to-day basis, but that was part of the job. Most of the other explorationists in Esso thought the Rover Boys lived a life of fun and adventure, and we agreed with the fun part—but we had a saying, "Having an adventure is when you screw things up." Even so, many strange and unforeseen things happened.

AFRICAN SNAKES IN THE BUSH

People have asked me about snakes in the bush, especially in Africa. Weren't we afraid of them? Did we ever see any? The answer is yes to both. We always prepared for danger ahead of time, and would read up on the dangerous flora and fauna before going out into the basin study areas. This included snakes. By learning about them you lose your fear, but not your caution.

Once in Guinea, West Africa, I was walking through some boulders out in the bush, when a big black cobra rose up next to me, hood spread and hissing. I froze. Eyeball to eyeball and about 2 feet apart, we looked at each other. But I knew that cobras are slow strikers, unlike rattlesnakes. The higher a cobra rises off the ground, the slower it strikes, because it has to swing its whole body, and you can slap it away with your hand. So I waited, not moving. After about 30 seconds or so, the cobra decided there was no danger, went down, and crawled off into the rocks. Afterwards you get the shakes, but that's always

Native guides used during 3-month geological foot safari in Guinea, West Africa. The spears are for protection against lions.

afterwards.

Were any of us ever bitten by snakes? I was, once, in the pants cuff. I was on a foot safari in West Africa in 1958. The trip was 3 months in duration and we covered about a thousand miles of trail, hunting for rock outcrops. I was accompanied by my interpreter, an old African who could speak Portuguese (to me) and Futa Fula to the locals. Following him were a half dozen African porters carrying our supplies on their heads—just like with Stanley and Livingstone. It was the dry season and I was walking down a trail looking off into the bush when I felt a thorn branch hook in my pants cuff. I took a couple more steps and it was still hooked on, so I kicked out my foot, and a 3-foot-long black cobra went sailing through the air. He hit the ground and crawled off. I knew that there was no way he could have bitten me through my pants cuff and boot, so I just kept on walking.

Behind me the Africans were pop-eyed. They had seen the snake rise up, hood spread, and strike me, and I had walked on, paying no attention, dragging this snake along. Then I kicked it to one side, hardly giving it a glance as I walked onward. They watched me the rest of the day as we trekked through the bush, to see when I would fall over dead. When we reached the next village they told everyone how the PATRON had been bitten by a cobra and didn't stop, didn't even look at the snake, and then walked all day and didn't get sick or die. They thought that I was immune to snakebite—that I had a powerful juju, or magic.

In the African bush, everyone knows what you are doing. In a couple of weeks the bush telegraph—talking drums—had spread the snake story far and wide. One day I was out looking at some rocks, and when I got back to the village in late afternoon there was a crowd under a big mango tree. One of the porters had been bitten by a snake. He was lying there, his eyes dilated and his skin grey and cold.

I asked a few questions to find out what kind of snake had bitten him. Did it have a thick body and a big triangular head? If so, it was a Gabon Viper, and the porter would soon be finished. I discerned that it probably had been a small black cobra. Nerve poison. If he were susceptible to it, he would already be dead. But this was Africa, and psychologically, the man believed he was going to die—as did the villagers around him. And as things stood, he *would* die, soon.

So, I stepped up and said, "No, this man will not die. I will give him some medicine against snakebite."

Old Selou, my interpreter, said, "Yes, it's true. The PATRON gets bitten by snakes all the time and he never dies." The story was improving with time. I went to my medicine kit and got out three "snakebite pills." I put these in his mouth and gave him a drink of water. Then I told the villagers to kill a chicken and make a stew, and I put some dehydrated vegetables and some more pills in the pot along with the chicken, a bunch of hot peppers, and whatever else I could find. The porter ate this mess—by then he was beginning to take an interest in what was going on. I gave him a few more magic snakebite pills and a big drink of water. Then I explained to all the crowd how the medicine was made by grinding up thousands of snakes in a big building to make the white powder and pills.

By then, everyone was listening to the story, so I went over to the porter and said, "When I first touched you, your skin was cold and your head was hurting. How do you feel now?" Everyone in the crowd went over and touched him and said he felt fine to them. The porter grinned and said, "I'm fine now, the magic works well."

I told them, "Tomorrow this man will get up and walk back to his own village." Everyone cheered. The next morning he got up and walked back home.

What did I give him? Bayer aspirin. As far as I ever heard, he recovered completely. The bad side of that event was that for the rest of the trip everyone in the villages expected me to cure them of their ills.

GEORGE AND THE LIONS

The most dangerous creatures in the African bush are microscopic, but no one is interested in hearing about them. Everyone wants to know about the big, dangerous animals, like the lions. Usually, if you are a hunter and know your way around in the bush, you can avoid the big animals. They are just as happy to avoid you, too, except maybe for crocodiles, which will come after you in the river. And we did have a few skirmishes with lions.

In 1959 we were working just north of a big game preserve in the Niger Republic of West Africa. We had shot a gazelle that morning and had tied it on the back of the jeep (we lived on a meat diet and shot 80 percent of what we ate). That afternoon we drove into the park. The Ranger at the entrance of the park sealed our guns, and then warned us about a pride of lions that hunted near where we planned to camp. A dog followed us from the ranger's hut. He was half-starved and after our gazelle. We set up camp and hung the gazelle up in a tree for safety. We scattered our cots around the clearing, and after cooking and eating gazelle steaks, we went to bed. In the firelight we could see eyes surrounding the camp: jackals, hyenas, and maybe the dog after our meat. At about three in the morning, George, our mechanic, woke up and heard the dog sniffing near him. He reached out in the dark next to his cot and grabbed a glowing branch from the now-dead fire, and belt-ed the dog a clout on the snout. It let out a yowl and ran off. George went back to sleep. Next morning he told us how he had bashed the dog with his club, so I went over to look at the tracks in the sand next to his cot. They were the tracks of a full-grown lion.

We wrote up the event in our next monthly report to headquarters, and added the following: "Now George thinks he is underpaid, so I have added one bottle of whiskey per week to his pay. He calls it courage water; and when the lions are roaring around camp at night he (and we) have a few snorts. Last night he confided in us, after downing a few, that lions didn't like the smell of whiskey. 'If you are skunk drunk,' George said, 'you smell bad, and they won't eat you. And even if they do, you don't care.' "

A SKIRMISH WITH THE TURKISH POLICE

Operating in foreign countries involved dealing both with the natives and with the local authorities. In southern Turkey, in 1955, we had more trouble with the police than with anyone else. Near the border, as soon as we drove into a town a policeman would come up to check our papers. The police had been trained to think all foreigners were spies, and they may have thought that they would show their superiors back in Ankara that they were really on the ball. Besides, nothing else was going on in their miserable villages full of dust, dirt

Water buck head, Chad.

and goat dung. Maybe he will get lucky and you really ARE a spy. As a reward they might re-post him to Ankara or Istanbul.

So with these dreams of glory in mind, a policeman would arrest us and then send off a telegram by Yildrim, which means like lightning, asking for confirmation. This "lightning" sometimes took a month to get to headquarters and back. Everyone was very spy-conscious along the Syrian border, and we kept getting arrested. To stop all this, I tried something new. The next time we were arrested I told the policeman, "We're here doing oil exploration. The country needs it and you are interfering by wasting my time. The Minister of the Interior back in Ankara has asked us to come to this country. He is a good friend of mine. He also runs the police. Now, my papers are in order. I know it and you know it. If you arrest me here and now, when I get back to Ankara, I'm going to tell my friend the Minister what a good job you're doing here—so good, in fact, that I think you should stay here for the rest of your life. "

The policeman started chain-smoking and pacing the floor, his dreams of being sent to Ankara or Istanbul fading. What if I did know the Interior Minister? But—what if I really was a spy? When he couldn't figure out what to do, that there was no way out for him, I gave him the answer. "I don't know why you are worried," I said, "because the police don't have jurisdiction over us, the Turkish Army is the one responsible for Americans." Of course, that was not true, but it sounded O.K. to him since most Americans there are military, and he was grabbing at straws. So the policeman sent us over to the nearest army camp. The colonel in charge (who probably had been trained in the United States) took one look at our papers and let us go.

This approach worked well, but after we had been sent back to this same colonel five times, he got sick of it. He called for a sergeant who, when under orders, could shoot anyone, and told him to get a machine pistol, a German Schmeiser, and ammo. When he came back the Colonel told him, "Sit in the back of this jeep. If anyone tries to stop them you say, 'I have been given orders to shoot anyone who stops this jeep.' Then you count to ten, and if they're still there, shoot them."

Off we went in the open jeep; no top, no windshield. Late in the afternoon we came into a town to buy bread, onions, and yogurt. A crowd gathered and there came a policeman who waved his hands in the air and told us to stop. I put on the brakes, stopping right in front of the policeman. Out of the corner of my eye I saw the sergeant unslinging his Schmeiser. He was smiling; he didn't like cops either. I ducked my head down under the steering wheel.

The Turkish sergeant, talking fast, said, "I've been given orders to shoot anyone who stops this jeep." And he started counting quickly to ten. I waited for the burst of gunfire, but—silence. I opened my eyes and looked up. There was no one—nothing but dust in the air, the whole crowd had vanished, including the policeman.

The word of our gun-toting Sergeant got around fast, and after that day we had no more trouble with the police. In fact, they became invisible. We'd drive into town and you'd see people running around and slamming their doors, like the Mongols were coming. We had a hard time buying supplies, but we had no more trouble with the police.

A BRUSH WITH SOME BANDITS

Our problems in Turkey were not always with the police or army; people who looked like villagers or shepherds by day could become smugglers or bandits by night. Robberies and kidnappings happened occasionally in the area. When we stopped in a village to buy food or

Hittite lion gate, Turkey.

supplies, we did not camp nearby, but drove 10 to 15 kilometers to some remote spot, usually on top of an easily defensible hill. We knew it might not be safe to stay too long in one place, since it gave bandits time to plan a raid. We traveled light and moved camp every 2 or 3 days. In this way our crew would come into an area with no advance notice, and do our work and move on again before potential trouble could arise. Of course, the bandits were not everywhere, but this system was very successful and using it we had few run-ins with nighttime visitors.

Sometimes, however, we would "do something stupid and have an adventure." In 1954, we were working along the Turkish/Syrian border, mapping some large anticlines that lay parallel to the frontier. All of the high points on both sides of the border were occupied by military or police Karakol stations that kept watch for unauthorized border crossings by bandits or smugglers. The intermittent streams that led across the border cut canyons or steep valleys through the hills. To avoid being picked up for questioning by border patrols, we located our camps in different places each night out of sight of the high ground, down in the canyons. No one could see our campfires from above. It was an ideal place to avoid unwanted attention from the army or the police. Unfortunately, the smugglers thought the same thing. Their route across the border, at night, turned out to be up these same valleys.

One night we were camped down in one of these canyons, just to one side of a boulder-strewn, dry stream bed. Supper was over and we were in our cots, sleeping under the desert sky. I was awakened at about 2 a.m. by the sound of horses' hooves on the rocks, coming up the valley. There was no moon, but starlight reflecting from white limestone around us showed a mass of horses and mules approaching. Our camp was visible—the darker vehicles, the still-glowing campfire. I put on my boots and stood up. Then I froze. I heard a dozen sharp clicks as rifle bolts locked in a round. Suddenly, it hit me what we had done. This deep gorge running across the border was a smuggling route, and we were camped right in it. The smugglers didn't know who we were. We could be a military ambush waiting for them. We were about to get our asses shot off.

It was time to act. Reaching down, I flicked on the single bare flashlight bulb we used as a camp light. It wasn't much, but it illuminated me with no weapon in hand. Then I did what we always do when things are bad—smile. Showing lots of teeth, I raised both hands in greeting. "Merhaba (welcome)," I called. Nothing happened. I threw some wood on the fire, and it blazed up. At the periphery of the firelight I could see about 30 tribesmen in a semi-circle around the camp, rifles raised. If this was an ambush, we would be the first to die.

Smiling some more, I said in Turkish, "There is coffee—American coffee—if you want some. But bring your own cups." No movement from the tribesmen, but behind them I could see a line of horses and mules with heavy packs passing on up the valley. By now the others in our camp were up. We started the Coleman stove and got the coffee

Dave Kingston examining Hittite soldiers carved in relief on rocks near Haymana, central Turkey.

water boiling. The smugglers were still there at the edge of the darkness. They seemed to be waiting for something.

After awhile, a figure rode up on a donkey and stopped. He was an old man with a white beard. The tribesmen began speaking to him. George, our Greek mechanic and interpreter, sidled over to me. "They're speaking in Kurdish," he said. "They're telling him what has happened. They think it's an ambush."

The old man abruptly rode into the firelight and dismounted. Standing on the ground next to his donkey, he was a smaller target. He was dressed in a long, ragged shirt, baggy pants, and a dirty old turban. He carried no rifle, which was a sign of his authority. I knew he was the leader. "Do you speak Kurd, Arab, or Turkish?" he asked.

We told him we spoke only Turkish. He seemed quite fearless and curious, and in Turkish asked us about the Coleman stove, the likes of which he had never seen. We gave him some Nescafe, which he drank with lots of sugar.

"Why do you camp here?" he inquired.

We explained that we were Americans looking for oil, and that we studied the rocks that were exposed in the walls of the canyon. We were also down here away from the police, I said, and I explained our problems with the authorities checking our papers. He looked at our clothes, especially our boots, which were American made, and our jeeps which were not military issue. He was checking our story against what he saw.

"Is there any 'petrol' around here?" he asked. I could see he was testing us to see what we knew.

"Yes," I told him. "Down the valley, near the border, there is a big oil seep that sometimes gets started burning by grassfires, and burns for a long time." He nodded his head. Did he believe us?

By then the loaded caravan had passed, and only a dozen rifle-toting smugglers remained behind. The old man mounted his donkey. He seemed hesitant about what to do next. I decided to help him out.

"Uncle," I said, "why do you have so many guns? Are you out hunting wild pigs?" The old man stiffened when he heard "guns," but then smiled when I got to the part about pigs. He gave a gap-toothed laugh for the first time, and his old eyes glittered.

"Yes," he said. "That's what we are doing—out pig hunting." And he laughed at the joke. The tribesmen in the darkness did not. "If anyone asks you, 'Did you see someone come up this valley at night?' you can say, 'Just some pig hunters.' " I nodded my head. "Go with God," he said, giving us the traditional farewell.

Then the old man turned his donkey and rode up the valley into the darkness. His armed escort followed him—all but two, who watched us until dawn. We never camped in those canyons again.

OIL PLAYS IN SOUTHEASTERN TURKEY

Our geologic studies in southeastern Turkey revealed that the area was a continuation of the Zagros, which is a broad foldbelt, rather than a basin as we had previously thought. Giant Middle-East-type anticlines were found along with oil seeps. The same or similar rock formations as had been identified in northern Iraq exist in southeastern Turkey, also. Previously discovered oil fields in Turkey, were small, however, occupying only the tops of the structures, and were definitely not of Middle-East size or caliber.

There is one significant difference in the stratigraphy between Turkey and Iraq, and it is in the Miocene Fars formation. In Iraq, the Fars is a salt series, forming a superseal overlying the reservoirs, as at Kirkuk. In northernmost Iraq at the Butmah oil field, the Fars formation is not salt, it is gypsum and red shale, so it does not seal or trap much oil. The oil at Butmah, for example, has been produced from below the Triassic salt—the next good seal down in the section. It is the same in Turkey; there, the Fars has no salt and does not act as a seal. Our conclusions, there-

Oil seep afire, Kirkuk.

fore, were that giant oil fields with Tertiary and Cretaceous reservoirs might not be found in Turkey, and we would have to locate structures below the Triassic salt, as in Butmah.

This turned out to be largely true in southeastern Turkey. For the rest of the Middle East (Iraq, Kuwait, Saudi Arabia, and Iran) we found that even though large structures, reservoirs, and source rocks exist, the presence of "superseals" is the key. In the Middle East, where "superseals" are present, either as evaporites or shale, there are giant oil or gas fields. Where superseals are lacking, the fields are small or nonexistent.

What were the results of our overseas efforts during all of those years? We had recommended a constant stream of plays and prospects to headquarters in New York. The company picked up and drilled some of these plays, along with farm-ins and prospects from other sources. Some represented Esso's big successes, while others were disappointing failures. The following are a few of the untold stories of how, in the 1950s and 1960s, Esso got into some big oil plays, and how they fared.

PARENTIS—THE FRENCH CONNECTION

After World War II, in the late 1940s, the French government opened North African Algeria to foreign companies for oil exploration. Esso sent an exploration team to do the reconnaissance geology and hydrocarbon evaluations. The basin areas were divided up into concession blocks and a geologic report was requested for each block desired, along with the dollar amount bid. Esso and the other foreign companies protested that they had never submitted technical reports before, but the French government decreed—"No report, no bid." The geologic reports on each block were reluctantly submitted. To no one's surprise, when the bids were opened, they all went to French companies, many of them newly formed for the occasion.

A few months later a second round of blocks was offered for bids. The conditions were the same, and again, the results were the same. Some of the foreign companies had learned their lesson and were successful by taking French partners. But Esso bid alone, and lost for the second time. The company complained that their reports were being used to give inexperienced French companies the selected acreage, but to no avail. Esso, by not taking a French company partner, had been shut out of Algeria.

Some time later, the French government invited Esso to bid on the Aquitaine basin in southern France. There was some resistance within the corporation about accepting this offer, because of the previous Algerian experience. But the financial terms were very good, and the area, though small, was interesting geologically. When the bids were opened, a French company had won the southern half of the basin, where the Pyrenean foldbelt anticlines were found. Esso, the only other bidder, was granted the northern half of the basin, which was a monocline that dipped gently southward from the outcropping Massif Central.

There was also a strip of unclaimed acreage in the middle of the basin near the coast, which was first offered to the French company. They refused because it was too far from the known structures of the mountain front. The government then informed Esso that we would also have to take the unwanted strip of acreage, along with the northern part of the basin.

Esso set up an office in Bordeaux and hired a bunch of energetic young French geologists to staff it. A seismic crew was sent over from the United States, and they unloaded their trucks and gear in Bordeaux. After hooking up their equipment (and draining a few flasks of Bordeaux Rouge), the happy crew set out for a test run. In later years, we always accused them of having headed south to be near the topless beaches of Biarritz. As they passed by an unused dirt road that happened to be on the unwanted acreage strip given to Esso, they decided to stop and do some test shooting. The sign on the road read "L'Etang

Parentis" (the lake of Parentis). The rest is history. The first line the crew shot went across the Parentis structure, which, when Esso drilled it, was found to contain reserves of more than 800,000,000 barrels of oil. After that Esso never complained about Algeria again.

ZELTEN—LUCK IN THE LIBYAN DESERT

Esso acquired another lucky acreage allotment, in Libya in the 1950s. At that time, what was then the Kingdom of Libya opened up the country for exploration. Esso sent a roving exploration team headed by Maurice "Abner" Wallace. Abner, a great field geologist and camp cook, used to swear he made the best "desserts" in the desert. (I can remember having one lemon pie cookoff competition with him using folding reflector ovens and ostrich eggs for meringue.) He also loved to work in the desert, and at the end of his field work in Libya, he had picked as No. 1 a Paleozoic prospect in the Fezzan near the Algerian border that was adjacent to some new French discoveries. Fortunately, Abner also liked the Sirte basin in Tripolitania to the east. When the blocks were assigned, Esso got their first pick—a block in

Rest stop in the Sahara. Wind-carved Cambrian sandstones in the Southern Hoggar, Niger Republic. Jeeps are outfitted for long-range desert reconnaissance with sand tires and sand tracks. Shelby Eddington on left, Dick Murphy and George Voutopoulos on right.

the Fezzan (later drilled and found to contain noncommercial amounts of oil), and another in the center of the Sirte basin. West of the Esso block in Sirte there was an unassigned sliver of acreage. Mobil, the adjoining blockholder to the west, refused the slice, and, as in the Aquitaine basin in France, it was given to Esso. Subsequent seismic on this sliver of land revealed a block-fault structure which, when drilled, proved to be the 4-billion-barrel Zelten oil field (now renamed Nasser). The lesson here is to always accept unwanted acreage in an unknown basin.

Libya was an interesting place to explore in the 1950s. Besides the general challenge of survival in the desert, there was the problem of the mine fields left over from World War II, which formed a belt along the coast and inland for about 60 miles. As the exploration in the basin progressed, veteran sappers from the German Afrika Korps and the British Eighth Army were employed to remove and explode the mines. These areas were then shot by seismic programs. More than one seismic "shot" triggered explosions of nearby unfound mines. Seismologists working in Libya in those days were well paid.

Clearing mine fields and shooting seismic were by far the biggest expenses in Libya. As a footnote, it was noted some years later that most of the large oil field structures in the Sirte basin can be seen on aerial photographs. A lot of money might have been saved had the oil operators completed their surface geological field work with air photo analysis, before they started in on the seismic.

Abner Wallace, who did the original field work in Libya, had this story to tell. Upon returning to Tripoli after some months of field work in the desert, he met a few friends from a Dutch ship that was in port. They all had dinner and drinks (quite a few of the latter), and then decided to row out to the ship anchored in the harbor to sample some "real Dutch Bols liquor." At 3 a.m. the captain announced over the intercom that they would soon be leaving port. Abner's Dutch friends saw him to the gangway which he unsteadily climbed down and stepped off into the dory. The ship quickly hoisted gangway and anchor and pulled away into the darkness. Unfortunately for Abner, there had been no dory there and he found himself floating in the middle of the shark-infested Tripoli harbor in the dark. He swam towards the shore and after a couple of hours it became daylight. Abner found himself floating off one of the harbor wharves, too tired to pull himself out of the water. He was finally spotted by a friendly native, who heaved him up on the dock, removing in payment Abner's watch, wallet, and shoes. But, Abner never complained, saying that he had learned a valuable lesson from the experience—"When you go swimming, always drink Dutch Bols liquor beforehand, and you won't sink."

Abner did have one complaint about Libya—that it ruined his record as the worst oil finder Esso ever had. He claimed that he had worked for 30 years in 34 different countries, none of which ever produced a drop of oil. Of course, everyone knew that Abner Wallace was one of the best geologists in the company, which was why he was sent on the most difficult assignments. But Abner liked it the other way around, and he always complained that Libya had ruined his reputation as "world's worst oil finder."

DR. MILLER'S FILES

In Europe, Esso had a long history of both good fortune and bad. Before World War II, the company had produced oil in both Hungary and Romania. When the war came, all expatriates escaped the Nazi-occupied zones except one company lawyer, an Englishman named Alex Miller. Dr. Miller, who spoke most European languages plus Japanese, collected the company exploration files for all of eastern Europe and hid them behind a fake stone wall in a mountain hut in Yugoslavia. He then found himself trapped behind enemy lines, so he went to Budapest, where a friendly paleontology professor gave him a job in his lab grinding rock slides. Miller thought himself safe until, in 1943, an SS officer picked him up with the words, "Come along, Dr. Miller. We thought you were an Esso spy and have been watching you for 2 years, but you've done nothing; so now we pick you up." Miller says he spent the rest of the war in a concentration camp practicing a weight-loss diet (starvation) and perfecting his German.

When the war was over, Miller gained his release and headed for the mountain hut in Yugoslavia. A civil war was in full swing, but he managed to retrieve Esso's exploration files, unharmed except for a few bullet holes, from the hut. He then smuggled the files to the coast and took them by fishing boat to Italy. There, in the newly reopened Esso office, Alex Miller was greeted by his friends as a man returned from the grave. The manager, wondering what to do with the technical files on Eastern Europe that Dr. Miller had rescued at such great personal sacrifice, wrote headquarters in New York and requested a decision. It looked as though Eastern Europe was going to be lost behind the Communist iron curtain forever. The decision was made and a letter came back from New York; the manager called Miller in and showed him the reply.

"Burn them," it read.

Afterwards, Alex Miller would tell us the story and laugh until the tears ran down his cheeks. "Four years in a concentration camp for nothing," he would laugh. But he was one tough cookie, and had been made even stronger by adversity. With the recent reopening of the Eastern European countries, perhaps Esso now wishes they had not burned Alex Miller's files.

ESSO AND SHELL IN EUROPE—A WARTIME MARRIAGE

The present-day Esso/Shell joint ventures in Europe had their origins in World War II. During the war years, exploration in Europe had shut down, and international companies scrambled to find new reserves in "safe" locations. Shell Oil (N.A.M.) got out of Holland one jump ahead of the German army and fled to London. Shell looked around and selected the island of Cuba as a possible exploration area, asking Standard Oil Company of New Jersey to go in with them on a joint venture. Esso was reluctant and Shell sweetened the deal by offering them half-interest in their Dutch concession (they had the whole country) that was under Nazi control—but for how long, no one knew. Esso eventually joined Shell in the Cuba venture without formalizing the terms.

The Cuba venture was unsuccessful. The company "La Estrella de Cuba" found nothing and was disbanded when the war ended in 1945. Shell went back to Holland and resumed exploration. A few years later, they made some oil discoveries along the German border. Management at Esso in New York suddenly recalled the pre-war deal and began a search for Shell's letter offering half interest in N.A.M. This was finally located and sent to Shell in The Hague, reminding them of their earlier offer for a 50-50 joint venture in Holland.

To Shell's credit, they stood by their word, even after all of the intervening time, and that is how the Esso/Shell joint ventures in Europe were initiated—first in Holland, then Germany, and finally in the North Sea.

We worked on the first Esso/Shell North Sea report in our European office in Geneva, Switzerland in 1962–64. It was the first time that we used the technique of constructing a regional cross section grid based on surface geology and wells, and tied together by long regional seismic lines. The technique identified about a dozen North Sea plays ranging in age from the Carboniferous to the Cretaceous. Our study missed only two main plays later found in the North Sea: the Cretaceous chalk (Ekofisk) and the Eocene deep water sands (Frigg).

The North Sea was originally considered to be an extension of the Dutch plays in Holland, with the Gronigen gas play number one on the list. We thought there would also be oil plays, if the basin turned out to be large enough. The Jurassic Dogger sands were high on the list as a potential oil play. In later years, many people have claimed credit for first recommending that play, but they are probably all wrong. We put the play in the first Esso/Shell report in 1962, but I cannot claim any credit because the play was explained to me in 1952 by a worldwide Jurassic expert, Dr. Hans Fribold, who worked for the Canadian Geological Survey.

THE BELEMNITE BATTLEFIELDS

In 1952, I was working in the front ranges of the Canadian Rockies using a horse pack outfit for transport. I met Dr. Fribold in Jasper, and he volunteered to show me a Jurassic section he had found in the nearby Snake Indian River gorge. He also did much more: he explained the regional Jurassic depositional patterns over the whole northern hemisphere. He pointed out that the Jurassic section is like a sandwich, with Liassic black shales at the bottom, sandstone in the middle (Dogger), and black shales with sands at the top (Malm). He explained that this same facies extends north to the Arctic, through Spitzbergen and down into Europe and Russia. The Dogger sandstones are filled with masses of fossil Belemnites. Fribold said, with his strong German accent, "It looks like dey had a great battle und killed each other off. I call dis sandstone der Belemnite Battlefields."

The Liassic black shales appeared to be a good source rock, the Dogger sands a reservoir, and the Malm both reservoir and seal. I asked Dr. Fribold where the Jurassic "sandwich" might be found in a basin worth exploring, and he replied, "Under the North Sea." For the year 1952, that was not a bad prediction. Of course, I never forgot his description of the "Belemnite Battlefields," and included the play in the first (1962) Esso/Shell North Sea Report. I never knew whether Hans Fribold lived to see his prediction come true.

NORTH YEMEN: ONE THAT GOT AWAY

Not all of our recommended plays translated into Esso oil discoveries; North Yemen is an example. In 1965 we made a study of East Africa that included plate location and paleodeposition maps. A newly published surface geologic map of Arabia provided most of the clues, along with measured stratigraphic sections along the Gulf of Aden and Stanvac-drilled wells in Somalia. Our restorations showed that the "oily" lagoonal facies of the Jurassic in the Persian Gulf basin does not extend southward along the Somali coastal basin, but does occur along the northwestern coast, along the Gulf of Aden. This

Minaret mosque, Abadez.

Jurassic lagoonal facies extends westward and crops out on the banks of the Blue Nile in Ethiopia.

The only connection of this western area with the Persian Gulf appeared to be through the small North Yemen basin, where several outcropping salt domes had been known since Roman times. We deduced from our regional paleodeposition maps that the salt was probably Hith equivalent Jurassic, making the existence of oil-source lagoonal shales a strong possibility. If there were salt domes, the basin would have to have some depth, making buried salt pillows with reservoir sands a distinct possibility. The whole play idea was quite simple and could really be worked out from the published geologic map of Arabia.

We recommended the play to headquarters as an exploration venture in 1965 and three or four times thereafter during the late 1960s and early 1970s. Each time, it was turned down for the same reason. Esso was part of ARAMCO, who had a 50-50 oil split deal with Saudi Arabia. Their contract also stated that if any ARAMCO partner (including Esso) offered better terms to anyone else (a 65-35% split, for example) they would have to offer the Saudis the same, and the ARAMCO take-home percentage might be changed. The ARAMCO partners were very sensitive to the 50-50 deal, and Esso found themselves unwilling to compete in many new venture areas around the globe where home country percentages were higher. Esso was forced to sit on the sidelines and watch other non-ARAMCO companies take over the new ventures in good-looking exploration areas such as Indonesia, the Gulf of Suez, West Africa, and parts of Europe and South America. By the time the ARAMCO 50-50 deal with the Saudis was finished, the really good new venture areas of the 1960s and 1970s had all been taken by other companies. We are often asked why Esso, which in the past had always been such a fierce competitor, suddenly, in the 1970s, ceased pushing for new exploration ventures. The ARAMCO deal was the main reason. Esso thought they had more oil than they could use. Why look for more?

The North Yemen play was finally acquired and drilled by others in the late 1970s. Esso then bought into the play after the discoveries were made (and the ARAMCO deal had expired). Perhaps Esso was disturbed because they had known about this play since 1965 and had let it slip through their fingers, but perhaps not.

CAÑO LIMON—CAUGHT BY THE MONSOON

Another near miss by Esso was Caño Limon in Colombia, South America. By the early 1970s, Esso believed that they had pretty well high-graded the available prospective basins in the free world, both on and offshore. Esso management decided that, to make sure that we had not missed anything significant, we should systematically evaluate *ALL* known basins in the world onshore and offshore (shallow water and deep), communist and non-communist, to see if there were any remaining undiscovered pearls.

We set upon this daunting task with a small group of explorationists (almost in the Rover Boy tradition). Using the new ideas of global plate tectonics, we reasoned that we could restore basins to their tectonic origins and then work out the subsequent depositional history. The results were far better than expected. The restorations were accurate and we developed a Global Basin Classification System that made it relatively easy to correctly predict unknown geologic events in basins. It was especially easy to work up similar plays in adjacent basins on the same or related plates.

One such restoration was the La Luna source shale and equivalents in the basins to the east of the Andes Mountains in South America. The eastern Venezuela basin had the richest of the La Luna source rocks for giant oil fields, including the Orinoco heavy oil belt. The basin was thought to be perhaps the single richest oily basin on earth. The updip, or shovel, end of the basin ought to be the locus of migrating oil, with the likely existence of more giant fields. The upper basin area went over a saddle at the Colombian border, where the next basin to the south, the Llanos-Barinas, began. The saddle ought to provide a regional trapping mechanism if localized structures could be found, and looked good on both sides of the border. Venezuela, however, was no longer available for foreign company exploration efforts, so we recommended the Colombian side.

Eventually, Esso obtained the permit areas from the Colombian government and began exploration, but quickly focused on the large foothill anticlines. A seismic grid was shot across the axis of the basin, and several regional lines were planned to extend eastward. The program had almost been completed when it was interrupted by the onset of the annual monsoon rains, which put the basin under water. The eastern reconnaissance lines were never shot, because it was deemed "too expensive" to bring the seismic crew back during the next dry season for just a few lines. The lines were shot later by Occidental Petroleum, and they crossed and identified the Caño Limon structure. That was a case of a first-class area not covered sufficiently by reconnaissance seismic methods to find out the real geology. By saving a little money, Esso lost a lot of oil.

SHOOTING THE ROADS IN SALAWATI

Another missed play was in the Salawati basin, at the western end of the island of New Guinea, in southeast Asia. The old Esso/Mobil combination of Stanvac had obtained this concession from the Dutch after World War II, before Indonesian independence. They were attracted to the area by large surface oil seeps coming out of Miocene limestone reefs.

Stanvac did some surface work, mapping several anticlines as well as shallow reef-like features. It was clear that more reefs might be found basinward to the west, and a seismic crew was sent in to shoot a program. The southwestern portion of the Salawati basin was rough country, with rugged limestone hills, and the roads followed the valleys between these hills. The seismic crew, on a tight budget and not wanting to cut lines across the rough topography, simply shot along the roads in this area. They found only a few small structures.

Stanvac drilled these very shallow reefs and found only heavy oil. They did better with the anticlines to the east and discovered two small oil fields. Then they dropped all non-producing Salawati basin acreage.

After Indonesia gained its independence and Irian Jaya (western New Guinea) became part of Indonesia, some geologists of Trend Oil Company looked over the area and did what Stanvac had not done.

They looked at the aerial photographs and were able to recognize reef trend topography from the surface outcrops. They mobilized a seismic crew, shot a seismic grid of straight lines across the southern limestone hills, and found the reefs—right under the hills. The old Stanvac program had missed the structures by following the roads. The roads winding between the hills had gone around each reef. Trend drilled these reefs and discovered large oil fields, including Walio.

Salawati was another case of major plays being missed because sufficient regional studies—in this case including aerial photos—were not done before seismic and drilling were begun.

LIFE COMES TO AN END, BUT THE ROAD GOES ON (TURKISH PROVERB)

During the 1950s and 1960s, most of the best-looking onshore basins worldwide were selected for exploration. Many contained oil. Many others were rejected as not looking prospective. In the late 1960s and 1970s, offshore exploration plays became the vogue. The onshore basins had been pretty well picked over by that time and not many obviously good-looking, accessible onshore areas were left. The field studies and overseas travel came to an end, and the roving assignments ceased. During the 1970s and early 1980s, the price of oil rose significantly, and the oil industry went on an exploration binge. Basins that we had repeatedly rejected in the past (we called them "dogs") were leased and drilled, mostly with negative results. The offshore basins on continental margins, which originally had been so highly regarded, were also explored with disappointing results. Only 15 to 16% of the continental margin basins were found to have commercial hydrocarbons. Today, only about one-third of the world's 650 sedimentary basins produce commercial oil or gas. The rest are dry, although some remain undrilled.

What are the prospects, on a global basis, of finding new fields and new plays, and where should we look? I believe there are undiscovered new plays in some of the sparsely explored basins and in many foldbelts around the world. Also, new plays will be found in presently producing basins, as long as people keep looking for them (note the Prairie du Chien in Michigan). What about new oil plays in Eastern Europe? Frankly, there are oil plays in Western Europe that haven't been drilled yet. Eastern Europe, having been politically, economically, and technologically "on hold" since World War II, should also have plays waiting to be discovered. Democracy should improve conditions for new drilling ventures in that part of the world by making advanced technology more available, and especially by providing an influx of new thinking and play ideas from western oil companies.

What did Exxon get out of us Rover Boys, during the several decades of our existence (besides entertaining material for the *Exploration Newsletter*)? Our contributions were many and diverse. Some of our reconnaissance field work led to acreage acquisition, and to drilling, a few of which resulted in major oil discoveries in Europe, North Africa, and South America. Very often our regional studies resulted in negative recommendations—to stay out of basins because they did not look "oily" from a geological standpoint.

Another of our contributions to Esso (Exxon) was the generation of a stream of new exploration ideas. Our assignments took us to all parts of the globe, to new and different basin types and oil prospects. In our travels, we saw numerous present-day sedimentation sites of great diversity. We crossed lakes and river deltas, salt flats and tropical swamps, arctic floodplains and carbonate reefs. The first-hand knowledge we gained of geomorphology was most useful in visualizing the paleogeography in ancient rocks and basins. When we returned to the U.S. after an overseas assignment, we would go to the research lab in Tulsa or Houston for debriefing. In turn, our research people would tell us of their latest geologic studies and literature evaluations. That exchange of ideas was very useful to both sides, and to the corporation as well, since it was an efficient means of collecting and testing new exploration ideas. The Exxon Global Basin Classification System, published by the AAPG in 1983, is one example of such an interchange.

Several important exploration principles were developed and used by the roving field parties. First, there was the value in getting the big geologic picture. Often, the answer to a specific problem was found not in the problem area, but in another part of the basin or in an adjoining area. Broad regional geologic reconnaissance is often overlooked today in basin evaluation. Second was the principle of data evaluation. First-order data constitute reality—that is, what exists in the basin area that you have actually seen and studied, such as outcrops, well cuttings and cores, and seismic tied into wells—the geologic truth. Truthful data will give you truthful answers and at the same time, fewer mistakes in basin assessment. A third principle we discovered was the danger of using technology without first-order knowledge and experience in an area. Incorrect interpretation of expensive seismic and geochemical data often results where the regional geologic history has not been correctly understood. Today, in some companies, "geological tourism" by management, combined with published maps and literature, often replaces regional studies by experienced geologists. The results can be unpleasant surprises during later drilling, and dry holes that could have been avoided.

EPILOG

People often ask me what has become of the other Rover Boys. Of our group, only one, Al Broun, is still with Exxon, and he is their manager in Trinidad. The rest of us have all retired from the corporation. Dick Murphy in London and Shelby Eddington in Madrid are both doing geological consulting. Pat Shannon and I are in Houston. Pat is doing air photo/satellite interpretation and I am consulting and teaching OCCI schools on worldwide basin evaluation. George Voutopoulos, our Greek mechanic, is retired and lives in Switzerland.

Of our predecessors from the 1940s group, only Giles MacEwen in Roswell, New Mexico, and Bruno Frasson, in Liechtenstein, still survive. Abner Wallace and Hank Marchesini both passed away in the 1960s. But during our years in the field, in many parts of the world and under sometimes difficult conditions, none of the Rover Boys were killed or suffered serious injury save one. Giles MacEwen was out hunting Cape Buffalo in Angola at night with a headlamp, and was accidentally shot through the stomach with a 12-gauge shotgun slug. He survived because, as Abner Wallace said, "Only the good die young."

Norway

Ekofisk: A Giant Discovery of Oil and Gas in the North Sea

By Ward W. Dunn

EKOFISK: THE NAME

Most of the people in the oil industry who were in any way involved in the exploration of the North Sea in the 1960s and 1970s would be familiar with the name Ekofisk, but those outside the oil industry deserve a little background. At that time, Phillips Petroleum Company was doing seismic work in the North Sea and had obtained production licenses on blocks of acreage in offshore Norway. We had a system to assign a letter of the alphabet to each block and to name the drillable structures after a fish or some other form of marine life. Everything worked out according to plan until we reached "E." After using Eel, we quickly ran out of fish and crustacean names. One of our geophysicists, using the imagination characteristic of his breed, coined the word "Ekofisk." You will never find the word in a Norwegian dictionary, and it is as mythical as the Norwegian troll. Literally, the English translation is "Echofish." The word "echo" came naturally to a geophysicist who had spent his life mapping echoes from subsurface sediments.

PHILLIPS ENTERS THE PICTURE

Phillips's involvement in the North Sea began in August 1962, when Paul Endacott, Vice Chairman of the Board of Directors, had just returned from a European vacation. At the end of a board meeting, he raised the question of whether the company had investigated the oil and gas possibilities of the North Sea. While in Europe, Endacott had noticed a drilling rig near Slochteren in Groningen Province, the Netherlands. Further inquiries had led him to believe that a major gas field had been discovered there in 1959. By 1962, ensuing drilling had confirmed the field's large size. Shell/ESSO, the operators, eventually developed a gas field that ranked with the largest in the world, having 66 trillion cubic feet (tcf) of gas reserves.

In the spring of 1962, I had been transferred to the Paris office as Manager of Exploration and Production for Phillips. I had been sent primarily to monitor the company's interest in concessions operated by the French in the Algerian Sahara. Secondarily, I was to look for exploration opportunities in Europe. When I arrived in Paris, we were involved in a major gas blowout at the discovery well at Gassi Touil in the Sahara. The well had caught fire, and we had some engineers advising the French. Finally, we called on Red Adair, whose company specialized in controlling blowouts, and he eventually brought the well under control. (A movie about Red Adair's life was made, starring John Wayne, and film footage from the well was featured.) With drilling suspended, the small exploration staff of three geologists were free to spend their time gathering data and compiling maps of possible exploration areas. We had also heard rumors of the Shell/ESSO discovery in the Netherlands, and had focused our attention on the North Sea and the surrounding countries, since it seemed to be the prime area to investigate.

We had already assembled a great deal of the published geological, geophysical, and legal data that were available in Europe, and we were in a good position for a more thorough investigation. About this time, we heard from Harry Brookby, Manager of the International Department, and Owen Thomas, Manager of Exploration and Production in our Bartlesville, Oklahoma office. We had top management approval to proceed swiftly with our investigation of the North Sea.

With that in mind, I scheduled a trip for October 1962 with Owen Thomas and Silvio Eha, Chief Geologist, Paris office. We planned to visit Germany, Holland, Denmark, and the United Kingdom. Our general objectives were to evaluate in depth the petroleum exploration possibilities; to recommend, if warranted, whether concessions should be acquired; and finally, if we had decided on a positive recommendation, to submit a plan for the acquisition, exploration, and exploitation of these concessions. Specifically, we would acquire all available pertinent geological and geophysical information and integrate it with data already acquired. In addition, we would obtain all the available land, legal, and tax law information bearing on petroleum exploration and production, and establish contacts who would be useful in obtaining further information as it developed and who would place Phillips in a favorable position for being selected to participate in any future concession awards.

Before we left on this trip, we had a meeting to review all the material we had gathered both in the U.S. and in Europe. We concluded that the area had a very favorable geologic column, with cap rocks, source rocks, and reservoir rocks all present. In addition, there were many oil and gas indications throughout the geological column from the Tertiary to the Lower Carboniferous. The major occurrence appeared to be in the Rotliegendes sandstone of Permian age, which was the reservoir in the Slochteren gas field in Holland.

By 1959, 40 years of previous exploration in northern Europe had attracted little worldwide attention to the North Sea. After intensive exploration during World War II, only small oil and gas fields had been found in the Carboniferous reservoirs of the East Midlands in the United Kingdom. Gas production had been established in Permian and Lower Triassic reservoirs, and onshore oil production in Upper Triassic, Jurassic, and Lower Cretaceous reservoirs in northwest Germany and in small oil and gas fields found in the Mesozoic in Holland.

Two factors dampened enthusiasm for the offshore area and slowed development. One was the lack of success in discovering fields that were major by world standards. The fields found to date onshore would have been noncommercial had they been offshore. The second, equally important factor, was the question of ownership beyond the three-mile limit. In 1959, the gas discovery at Slochteren in Holland changed all this. The discovery, combined with the increasing demand of the European energy market and the need to diversify supplies to this market, was the catalyst that initiated the offshore search.

Also at this time, the concession situation was being legalized for waters beyond the territorial three-mile limit. In 1958, the Geneva Convention on the Continental Shelf had established the ground rules (median-line principle) for the division of these offshore areas between countries. This convention, however, required ratification by 22 countries—which was not reached until 1964.

EARLY NEGOTIATIONS

After reflecting on this review, we decided we had a solid framework but that there were gaps to be filled in. We first attended a meeting of the DGMK (German Society for Petroleum Science and Coal Chemistry) in Karlsruhe, West Germany, where we made a number of useful contacts and heard some technical papers of value to our search.

We then proceeded to Bonn, West Germany to a meeting at the Federal Ministry for Economy. The title situation for offshore Germany was still vague. So far, only 14 countries had approved the 1958 Geneva Convention rules, of the 22 countries required for ratification, and Germany was not among them. Also, the decision had not yet been made on whether the offshore area that could fall under German jurisdiction would be under state (Lander) or federal domain. There were no predictions on when the settlement would be decided.

About that time, we again reviewed the situation and concluded that if and when the North Sea was divided among the bordering countries, Norway would have a large share if she was allowed by neighboring countries to disregard the 610-meter- (2000-foot-) deep, 32-kilometer- (20-mile-) wide Norwegian trench, which was a contradiction under the definition of "Continental Shelf" at that time. We then arranged a meeting with Mr. Bue Brun, the Commercial Attache at the Norwegian Embassy in Bonn. He was not aware that areas in the North Sea over which Norway would probably have jurisdiction, based on the Geneva Convention, might contain oil and gas possibilities. He was receptive and advised us to contact former U.N. Secretary Trygve Lie, who was at the time chairman of a committee composed of a number of cabinet ministers evaluating new investment opportunities in Norway.

Continuing our trip in Germany, we met with a number of German oil company executives and pursued the possibilities of joint ventures in the North Sea. We followed this with meetings in Hanover with the Federal Geological and Geophysical Service, with other state administration bodies, and with PRAKLA, a well-known German geophysical company. We were shown a great number of maps and cross sections of the North Sea and we focused on the fact that the geological formations definitely thickened northward. This was most apparent in the Tertiary, or youngest, section. This alerted us to additional oil and gas possibilities above the Permian Zechstein, and definitely increased our interest. It is a statistical fact that the greatest majority of oil found in the world has come from these younger sediments. The other feature notable to us was the abundance of salt deposits causing many major structural traps and seals. We had not expected to be shown as much data as we saw, and we were agreeably surprised. Whereas we previously had expected that favorable structures existed offshore, we now knew that an abundance of these structures did exist in a sedimentary section that had all the most desirable aspects for oil and gas accumulations. All this existed in Western Europe, an area of great economic expansion. We were becoming more excited as our journey progressed.

During a trip to Clausthal-Zellerfeld, West Germany, the administrative and executive branch for petroleum and mining matters, we were shown a map from a Dutch newspaper. The map indicated its view on how the North Sea should be divided between the bordering countries, based on the median-line principle that had evolved from the Geneva

Convention conferences. We felt, however, that there would be tough bargaining between the countries. For example, Germany's concave coastline penalized her in acquiring a fair share. Norway stood to acquire a large share if it could negotiate a common line with the U.K. and Denmark, without regard to the Norwegian Trench, which, by some interpretations of the Geneva Convention's definition of the continental shelf, would limit Norway's jurisdiction. We learned that a number of oil companies had already talked to the German authorities, that the French, BP, Shell, and Brigitta had done offshore seismic work, and that PanAm (AMOCO) had made an offer of $2.5 million worth of seismic work in exchange for an assurance of receiving a specific concession area once the jurisdictional matters were settled. We were extremely happy to have accumulated so much technical and political data from a source from which we had expected little.

Continuing on to Copenhagen, Denmark, we had meetings with the A.P. Moller firm. We learned that A.P. Moller had been given a 50-year concession on all of Denmark, including the territorial waters, and priority on any offshore area that might come under Danish jurisdiction. They were already far along on negotiations with Gulf and Shell as partners to form the Danish Underground Consortium. Obviously, we were too late, but they promised to consider us if they sought additional partners in the future. During our visit to Copenhagen, we took the time to write Ambassador Trygve Lie in Norway to request a meeting with him sometime after October 26.

We then visited The Hague in Holland where we had meetings scheduled with the U.S. Embassy. They were poorly informed about the North Sea exploration scene, but supplied us with contacts in the Department of Mines, The Netherlands Industrial Institute, the Ministry of Economic Affairs, and the Geological Service located in Haarlem. We followed up these contacts and they mostly confirmed what we had already learned. Shell had the inside track because The Netherlands is their worldwide headquarters and they were the operator for the Slochteren gas discovery, under Napoleonic laws. Under these laws, for example, in mining one had to find the mineral and then obtain a concession. This was not practical in the oil business and would cause major problems. The winds of change, however, were beginning to penetrate the bureaucracies with the aim of revising the existing laws to terms that would encourage oil exploration. We again obtained a great deal of important geological and legal data.

Next, we flew to London. Our first meeting was with Angus Beckett, the Undersecretary of the Gas Division of the Ministry of Power. This first contact was a lucky one, because Angus later became the most influential government official in forming and executing a policy for exploration and exploitation in the United Kingdom portion of the North Sea. We also met with his counterpart, Mr. Charles Thorley, Undersecretary of the Petroleum Division. The United Kingdom at that time allowed exploration (seismic) work to be undertaken beyond the three-mile limit, but they emphasized that no licenses could be granted at the time and the work would in no way constitute a right to obtain concessions nor would it make any difference whatsoever in the awarding of concessions that would be available in the future. We knew of the seismic work presently underway by BP, Shell, and ESSO, and were assured by other remarks that the British government intended to keep an open-door policy. It was reassuring to learn this, and although we still thought BP, Shell, and ESSO would all have strong positions in acquiring concessions, there appeared to be a large enough area that we could compete. It was also good to hear that the government was on top of the situation and was in the process of revising their petroleum rules and regulations and keeping a close eye on the Geneva Convention. We had an overall favorable impression and felt that the U.K.'s offshore oil and gas possibilities were highly attractive. Roughly half of the North Sea area would come under Her Majesty's Government's jurisdiction. The geology appeared very promising and the economic and marketing conditions were as good as, or even better than, those in the rest of Europe. We also believed that the future concession terms would be favorable.

In the meantime, we had received word from Norway that a meeting had been scheduled with Trygve Lie on October 27th, so we set out to keep it. To summarize, our feelings at this point were:

West Germany—A portion of the possible area appeared favorable geologically. Offshore regulations and policy were not yet formed. A possible large consortium would be established. One could become a member of this or enter a partnership with a German company. The biggest negative factor was that this would probably be only a small interest, in the smallest geographical area in the North Sea.

Denmark—It was too late in Denmark; the consortium was already formed. The geology appeared favorable but the chance of acquiring an interest appeared to be slim to none.

Netherlands—The geology was very favorable, but it was governed by a very poor petroleum law. In addition, there was formidable competition from Shell, which had a dominant role in the country and was one of the world's largest oil companies. Their joint venture at Slochteren with ESSO made them even more formidable. Still, it was a possibility if Holland followed the U.K. lead to divide the area into blocks.

United Kingdom—This area was very attractive geologically, with a potentially large area for concession awards. An already favorable petroleum law was being revised to take care of the North Sea situation.

ORIGINAL NORWEGIAN CONTRACTS

Our meetings with Trygve Lie and a few other Norwegian officials on October 27 and again on October 29 were most cordial, but they were incredulous that we believed there were oil and gas possibilities in what would likely be Norway's portion of the North Sea. You can understand their emotions when you read this contemporary statement by the Norges Geologiske Undersokelse: "The possibility of coal, oil or sulfur along the Norwegian coast should be disregarded."

The only problem was the 610-meter (2000-foot) deep, 32-kilometer (20- mile) wide Norwegian Trench fringing the coastline and separating Norway from the relatively shallow continental shelf area of the North Sea. A strict interpretation of the Geneva Convention would confine Norway's jurisdiction to the east side of the trench. During the negotiations at the Geneva Convention, Norway had paid little attention to the oil and gas possibilities of the North Sea and had concentrated on the fishing rights. The Norwegians were pleased to have this possibility called to their attention. They would plan to press for their fair share of any territorial division of the North Sea. Norway could easily end up with a share equal to nine million acres, second only to the U.K. The potential area would extend farther north than the U.K.'s but would be limited by increasing water depth and adverse weather conditions. The North Sea's greatest depth west of the Norwegian Trench is about 67 meters (220 feet), which at that time was the approximate limit for offshore producing operations. They told us that we were the first oil company to contact them, and we definitely felt we were in on the ground floor. Because of the seismic work we had seen in Germany we felt sure that there were excellent possibilities for structures and sediments that could contain oil and gas.

Because of the situation, we felt we should get a proposal in writing as a basis for negotiations. We wrote a letter to Trygve Lie simply stating that Phillips Petroleum Company was ready to commit $1,000,000 in seismic work in the spring of 1963 in exchange for a concession on the area to be explored. In the meantime, the Norwegian government would issue laws and regulations pertaining to exploration and exploitation of oil and gas. We then returned to Paris and received a reply on November 2, 1962, in which Trygve Lie stated he had turned our letter over to the Ministry of Industry, who would reply in due course.

CONCESSION DEVELOPMENTS IN NORWAY

In June 1963 Phillips and other oil companies obtained approval to do seismic work in offshore Norway, and we proceeded to do so. We also did about $600,000 worth of seismic work in offshore England, West Germany, and The Netherlands, and ran two east–west lines into Danish waters. As a result of these lines, we were immediately informed by the Danish Government that we were trespassing and would have to turn over the seismic records to them or face fines or possible jail terms. We speedily complied with their demands.

In the meantime, Norway was not wasting much time in asserting jurisdiction over what they now believed to be their portion of the North Sea. On May 31, 1963, they issued a royal decree declaring the submarine areas outside the coast of the Kingdom of Norway to be subject to Norwegian Sovereignty for exploration and exploitation of natural resources. This was followed on June 21st by an act vesting the rights of these natural resources in the state and giving the king the authority to issue regulations concerning said exploration and exploitation. As a result of these acts, a Continental Shelf Committee was appointed to prepare the necessary legislation. The Chairman was Jens Evensen, then Head of the Legal Department of the Ministry of Foreign Affairs, who is now serving as one of the judges on the World Court in The Hague. The committee included representatives from the Geological Survey, Geophysical Department, University of Bergen, Fishing Authority, Ministry of Industry, and Ministry of Justice. The Secretary was Leif Terje Loddersol, who was later President of Den Norske Creditbank in Oslo. In addition to deciding on the proper laws to govern this oil and gas exploration and exploitation, they negotiated with the United Kingdom, Sweden, and Denmark to establish a median line border between Norway and these countries. They were successful in all counts, gathering information on petroleum legislation and taxes from governments of oil-producing countries and from the oil companies themselves. The boundary lines were established with the U.K. and Denmark by 1965, on the median-line principle, despite the arguable problems of the Norwegian Trench.

PHILLIPS EXPLORATION ACTIVITY IN THE NORTH SEA

In the spring of 1963, we had a change of management in Phillips's International Department. Ed Van der Bark replaced Harry Brookby as Manager, and we had a new selling job for North Sea exploration. However, we received the needed support to push ahead with our planned 1963 seismic work and to make seismic trades. In addition, we were allowed to increase our geological, geophysical, and negotiations staff. Because of the size of the area involved, the exploration and development costs would be very expensive and we decided to reduce this expense and lower the risk by bringing in partners.

We brought in two European partners: Petrofina, a private Belgian company, in 1965, and later we added AGIP, the Italian national oil company. This was done to reduce our risk in a large, untried, and potentially expensive area. In addition, because of the political advantage we tried to bring in local partners. We were successful in this in the United Kingdom, but we were unsuccessful in joining with a Norwegian group made up primarily of ship brokers and ship owners. Their terms were too exigent and they finally signed an agreement with AMOCO.

To make our group more attractive to the governments involved, we contracted to have drilling rigs constructed by U.K. and Norwegian firms. A jack-up drilling rig christened the *North Star* was constructed at the John Brown shipyard on the River Clyde outside Glasgow, the same yard where the *Queen Mary* and more recently the *Queen Elizabeth II* had been constructed. In Norway we contracted with Offshore Drilling and Exploration Company to have a drilling rig constructed there. This was a semi-submersible rig, which could operate in deeper and rougher water than a jack-up. It was built partially in Sweden and partially in Norway, and was finally assembled, rigged, and christened *Ocean Viking* by my wife at Akers shipyard in Oslo fjord in September 1966.

CONCESSION TERMS—NORTH SEA

The other development of importance in the period 1963 to 1968 in the United Kingdom, Norway, and Holland was the promulgation of laws, rules, and regulations to govern the exploration and exploitation activity. Although the regulations generally covered block size, rental costs, terms, royalty, and the like, they differed in specifics because of the difference in risks and geological attractiveness involved and the desire of the countries to encourage the greatest number of applications and to negotiate the most realistic work programs. A U.K. license block covered about 233 square kilometers (90 square miles or 57,600 acres), a Dutch block covered 415 square kilometers (160 square miles or 102,400 acres), and a Norwegian block covered 544 square kilometers (210 square miles or 134,000 acres). Other significant differences included the royalty (U.K.—12-1/2%, Norway—10%). This governmental process was accomplished expeditiously, with the U.K. offer-

Mary Dunn christening the *Ocean Viking,* **Oslo, Norway, 1966.**

ing licenses on concession blocks in 1964, Norway in 1965, and Holland in 1968. The Phillips Group was successful in acquiring licenses in all of these countries.

In 1965, because of the increased activity in the United Kingdom, my staff and I were transferred to London to run the European exploration and production operations from there. We also stationed representatives in small offices in The Hague in the Netherlands, in Hanover, West Germany, and in Oslo, Norway. These offices reported to me in London. The Algerian operations were being nationalized and we still had a single representative in Paris to handle these negotiations.

MAJOR GAS DISCOVERIES—U.K. NORTH SEA

Exploration drilling in the United Kingdom was almost immediately successful with BP discovering the West Sole gas field in 1965, a year after licenses were first awarded. Other major gas fields were quickly found, including Leman Bank (Shell/ESSO), Indefatigable (AMOCO/British Gas Council) and Hewett (Phillips Group/ARPET) in 1966, and Viking (CONOCO) in 1969. These offshore gas fields were tied by pipelines into three terminals on the east coast of England with an initial capacity of 2 billion cubic feet of gas per day (BCFGD) and a target capacity of 4.5 BCFGD by 1975.

During this period of outstanding success in the discovery of gas in the southern portion of the North Sea in the U.K. and in The Netherlands, operations continued in Germany without success after 17 consecutive dry holes. In The Netherlands, offshore exploratory drilling began in 1968 shortly after concessions were awarded, and gas was discovered in December. In Denmark, offshore exploratory drilling started in 1966 and significant oil and gas shows were encountered in sediments of Danian and Late Cretaceous age. Production commenced in 1972 from the Dan Field at the rate of 4000 to 5000 barrels of oil per day. This development had been delayed by the Danish boundary dispute with Germany and the low productivity of the wells, which had made commercial production questionable.

INITIAL EXPLORATION IN NORWAY

In the meantime in Norway, after completing seismic work in 1963 and consulting with the government on forthcoming rules, regulations, and laws concerning future exploration and exploitation, in 1963 and 1964 we prepared a recommendation to Phillips's management to acquire licenses. The geological structure and sediments were most favorable for oil accumulation. It was evident that the Permian (the producing reservoir at Slochteren) extended beneath the North Sea, and the Mesozoic and Tertiary sediments, which outcropped around the southern margin, also extended beneath it. Geophysical data and onshore drilling supported this theory. From a structural point of view, the general effects of the Hercynian, Kimmerian, and Alpine orogenies were well known, and the structural and cap rock potential of the thick Permian and Triassic evaporites could also be deduced from the onshore geology. The pre-Carboniferous stratigraphy was known from the surrounding land areas, but due to metamorphism resulting from the Caledonian orogeny, these rocks offered little incentive to the oil explorer. Of more interest were the deposits of the Lower Permian, the Lower Triassic and the Upper Jurassic–Lower Cretaceous, and possibly the Tertiary, as likely reservoir rocks. Sediments of Carboniferous, Permian, Lower Jurassic, and Tertiary age offered promise as source rocks. A wide variety of features providing hydrocarbon entrapment were present: salt-controlled traps, fault traps, stratigraphic traps, unconformity traps, anticlines, domes, etc. All these classic traps were present onshore and were indicated offshore by the new seismic work.

The seismic work in 1963 and 1964 not only confirmed our thinking, it made us even more enthusiastic. In addition, the concession terms were reasonable in the United Kingdom, Norway, and Holland, the countries in which we were most interested. The market in Europe was substantial and increasing yearly. Therefore, the only major drawback we could foresee was the high cost of exploitation and exploration because of the weather and water depth. North Sea weather was infamous for its severity, and when combined with water depths of 61 meters (200 feet) and possible operations 322 kilometers (200 miles) offshore, it would increase the costs significantly. This was all explained to our management, but the convincing argument was displayed to senior management on the floor of the company gymnasium where the AAU championship Phillips 66 basketball team had played. The seismic cross sections were laid out on the floor and explained by our geologists and geophysicists. Management became convinced that we had a viable and potentially economic project. With this important endorsement, we had more or less "carte blanche" to pursue the acquisition of concessions in the North Sea, with special emphasis on the United Kingdom, Norway, and The Netherlands.

FIRST CONCESSION AWARDS—NORWAY

The applications were opened for the first Norwegian Production Licenses in June 1965 and were granted on August 18, 1965. The Phillips Group had applied for a total of 28 blocks, with Block 2/7 (Eldfisk and Edda) given first priority. Block 2/4 (Ekofisk, West Ekofisk, and large portions of Tor and Albreskjell) was further down the list. In awarding licenses, strong consideration was given to the applicant's financial strength and practical experience in oil exploration and exploitation. Consideration was also given to the extent the applicants had, and would contribute to, strengthening Norway's economy through marketing, construction of refineries, utilization of Norwegian ships, and the like. Production licenses were awarded for a total of 78 blocks. The Phillips Group was awarded three production licenses containing a total of 11 blocks. An interesting sidelight is the fact that our group was the only applicant for Block 2/4. There were many applicants for Block 2/5, contiguous to the east, which contained a much larger structure. The AMOCO/NOCO Group was the successful applicant, but this attractive structure turned out to be mostly nonproductive because of subtle stratigraphic changes not apparent from seismic work.

Although Phillips had already set up an office in Oslo primarily for government contacts, an operating base was needed on the southern coast. Ed Crump was transferred from Phillips operations in Tripoli, Libya, as Drilling and Production Superintendent for this purpose. He arrived in Stavanger in November 1965 and we made the decision to place the shore base on the southwest coast outside Stavanger at Dusavik. The mayor of Stavanger, Arne Retterdal, and a shipowner, Torolf Smedvig, were instrumental in locating the base there. Early in 1966, Ed Jobin, a drilling and production engineer, was transferred to Oslo as Manager. Phillips expanded the number of personnel and we proceeded to rig up the *Ocean Viking*, construct the shore base at Dusavik, ship in material for drilling operations, line up other supplies, and firm up a drilling program. With the awarding of the three production licenses, the Phillips Group took on a firm commitment to drill five wells on separate structures and at least one on each license over a period of six years. When we analyzed the areas covered by our blocks, we decided that it might be a good idea to spread our risk by perhaps trading a percentage of our blocks with another operator for a like percentage in their blocks. This was arranged in 1966 with the French group (Petronord), which had as a partner an influential Norwegian chemical company (Norske Hydro). They had been awarded three licenses for a total of 12 blocks. The equity interests then became (Phillips Petroleum Company—36.96%, Petrofina—30%, AGIP—13.04%, Petronord—20%).

EARLY DRILLING—NORWAY

ESSO was the first company to drill a well on Norwegian licenses. The rig *Ocean Traveler* was towed from New Orleans to Stavanger and ESSO spudded their well on Block 8/3 on July 19, 1966. Their next well was drilled in Block 25/11 and had shows of oil and gas. By the time they completed the well, the North Sea weather had damaged the rig enough to suspend operations for four months. Experience gained from the *Ocean Traveler* by the contractor ODECO was used to modify the *Ocean Viking*, which resulted in further delays for us.

During the period 1965 to 1968, there had been considerable success in the southern portion of the North Sea, with major gas discoveries in the United Kingdom. These early successes had spurred a great deal of drilling activity in that area. However, this drilling led to fewer and smaller gas discoveries, with the exception of the Viking field, and the early aura of optimism gave way to pessimism. Drilling activity had sharply decreased by 1968. The only significant oil show was encountered in one well drilled in this area, off the Norfolk coast, which tested at the rate of 400 barrels of oil per day—not enough to sustain commercial production. At the price the British Gas Council was willing to pay for gas, only large fields would be attractive commercially. The expense of putting a small field on production was definitely not a commercial proposition. The first enthusiasm was fast disappearing in North Sea exploration.

At this low point, however, ESSO, the AMOCO Group, and the Phillips Group pushed on with their exploration plans in Norway. Because of ESSO's problems with the severe weather with a rig very similar to ours, and AMOCO's problem with a converted whaling ship, we decided on a shakedown in shallower waters closer to shore. Thus the *Ocean Viking* left Oslo March 17, 1967, under tow to United Kingdom waters to drill. As a result of our experience drilling the first well, further modifications were undertaken in Stockport, England. Our drilling strategy in Norway was to drill the most interesting and potentially commercial structures, but we were also using caution because of the severe weather. The final decision was to move the rig to the northern licenses, where we would have the shallowest water, and then to move progressively farther south to deeper water and greater distance from shore. The wells would also test a thicker and thicker geological section as we progressed southward. In the north, we would be operating in less than 200 feet of water and about 130 miles offshore. The first well was spudded July 14, 1967 and completed as a dry hole in September, at which time the rig was subcontracted to Petronord.

During 1968, Phillips underwent a complete reorganization of its international activities, with the emphasis on decentralization. Phillips Europe-Africa was set up and headquartered in Brussels, Belgium, and was responsible for all of Phillips's activities in that geographical area. C.J. Silas was appointed President and I became Vice President of Exploration and Production at that location. This added to my duties because of exploration and production activities in Egypt, Nigeria, and other countries along with our continuing activities in the North Sea.

COD DISCOVERY—NORWAY, 1968

On February 26, 1968, the *Ocean Viking* was returned to the Phillips Norwegian Group and commenced drilling a well on the "Cod" prospect in Block 7/11. In June this well was drilled to 13,124 feet and tested at the rate of 40 million cubic feet of gas and 2000 barrels of 51° condensate per day. This discovery caused an enormous amount of attention because of the lack of new drilling in the southern part of the North Sea and the failure so far to come up with any significant oil or gas discoveries in the northern portion. A second, or confirmation, well was completed in October 1968 and tested at about the same rates

Ocean Viking **drilling in the Norwegian North Sea, 1968.**

as the discovery well. The major problem of putting the Cod Field on production lay in crossing the Norwegian Trench by pipeline to Norway. Studies were made and the conclusion was reached that the cost would be too high for the anticipated amount of production. During 1969, the Phillips Group operated and/or participated in a further six dry holes with no significant shows of hydrocarbons. The Group had spent about $12 million in seismic work and drilling, and our management was becoming restless. More than 30 wells had been drilled on the Norwegian acreage by ESSO, the AMOCO Group, the Phillips Group, Murphy Oil, and Syracuse, with little to show for it except "Cod," which was a teaser. At that time, Phillips was making the mid-year budget review and decided that Norway would be a good place to reduce expenditures, so we were told to subcontract the remaining time on the *Ocean Viking* contract. We tried, but were unsuccessful because of the pessimism and the low rate of drilling in the North Sea at that time. In fact, there were only three rigs active in Norway early in the fall of 1969.

Back in March 1969, the Phillips/ARPET United Kingdom Groups had signed a gas contract on the Hewett gas field in the United Kingdom. This was roundly criticized for setting too low a price. Once again, though, we were trying to protect our position by getting a cash flow as soon as possible in order to be in a profitable position. Because the British Gas Council had previously agreed to a much higher price for British Petroleum's North Sole gas, there was some justification for the complaints. Another problem was the large amount of gas reserves found by major companies such as the Shell/ESSO group and the AMOCO/British Gas Council Group. We felt that if we didn't sign immediately, we would lose our place in the queue and it would be many years before we would see a cash flow for further exploration projects.

With all this in mind, the next well offshore Norway, 2/4-1, was presented to management for approval and was resoundingly turned down although the Phillips Group had not met the total drilling commitment with the Norwegian Government on our licenses. Having been unsuccessful in subcontracting the rig, in August 1969 we went back to management one more time, with Owen Thomas making the presentation before the management. Our contract on the *Ocean Viking* rig was still in effect, so we had to pay a minimum charge even if we stacked it. Finally, Phillips's management approved the well and we moved the *Ocean Viking* to Norwegian Block 2/4.

An unfortunate story has circulated for years—that the Norwegian government forced us to drill this well. They had nothing to do with our decision, since under the license obligations we had until 1971 to drill our wells. We simply decided to keep drilling on a legitimate prospect rather than pay for an idle rig.

EKOFISK DISCOVERY—NORWAY, 1969

On August 21, 1969, the well 2/4-1 was spudded in the middle of the North Sea on the Ekofisk prospect, a north–south-trending anticline about 322 kilometers (200 miles) from shore and in a water depth of 67 meters (220 feet). We had traces of gas at a depth of 900 meters (2952 feet), and two days later, at a depth of 1663 meters (5452 feet), we hit a gas pocket, lost circulation, and oil began coming over the shale shaker. After many lost circulation and mechanical problems, the well was junked and abandoned, and the rig moved one kilometer south-southeast, to spud a new well, the 2/4–1 AX, on September 18, 1969. Surprisingly, there were no significant shows in this well in the

Ekofisk complex, 1989. Approximately one mile top to bottom.

Miocene, where the extensive oil and gas shows had been encountered in the abandoned well. This was a disappointment, but other major objectives lay below. Drilling continued to 3084 meters (10,109 feet), where oil and gas shows began to appear. Oil and gas shows continued to 3195 meters (10,473 feet), despite lost circulation problems, and we prepared to run a production test over this interval of 111 meters (364 feet).

Numerous serious weather delays occurred, including the rig being blown off location. During this time, we ran three tests that were unsuccessful due to mechanical problems. A fourth test was successful, however, and during a six-hour period the well produced 1071 barrels of oil (a rate of 4284 BOPD) and produced gas at the rate of 5.9 million cubic feet per day (MMCFGD). These measurements were highly inaccurate because of the brief testing period and the lack of separation equipment. By Christmas 1969, the well was secured and the rig moved off. We felt the well really had greater productivity than the fourth test had indicated, and we were especially impressed with the thick oil column. It was, however, a most frustrating time because of our inability to get valid test data.

Just how big a field had we found? From the test results, formation cores, and from our seismic work, we knew we had a north–south-trending anticline measuring 7.5 miles north–south and 4.5 miles east–west. It contained more than 55 square kilometers (13,625 acres) under closure. The reservoir was limestone and chalk, with over 600 feet of oil column and a net pay of nearly 400 feet. The porosity was 30% and the permeability was 10–12 md. The gas-oil ratio (GOR) was 1600:1, and the oil gravity was 36° API. It was obvious that we had discovered a giant field, providing the reservoir was homogeneous over the area of closure and had a good recovery factor. Recoverable oil reserves could be estimated at approximately one billion barrels. Six fields—Ekofisk, West Ekofisk, Tor, Albuskjell, Edda and Eldfisk—were eventually discovered and produced in the Ekofisk area. These plus the Cod field, located 81 kilometers (50 miles) to the northwest, were tied together to make up the seven fields referred to as the Ekofisk Project.

THE EKOFISK PRODUCTION PROJECT

The Ekofisk Project was developed in four phases, primarily in the Bartlesville office under the direction of Bill Boyce, Manager of Drilling and Production International. Phase I was to get the Ekofisk Field on early production in order to obtain production and reservoir performance information, prior to committing to the tremendous investments required for the overall project. This project would also recoup some of our initial exploration expenses by creating a cash flow. This phase was a plan for temporary production with conventional equipment that would be used in an unusual manner. We completed the original well, 2/4-1AX, and three other wells that were drilled in 1970, using subsea wellheads on the sea floor. These four wells were then connected by flow lines to a former jack-up drilling rig that had been outfitted with production equipment. After separating the oil, gas, and water on the platform, the oil was pumped by sea-floor loading lines to mooring buoys and from these buoys directly into a tanker for shipment. Bad weather allowed us to load tankers only 82% of the time. The gas was flared. This phase cost about $28.5 million and was completed July 7, 1971—about 20 months after the first significant oil and gas shows in the discovery well. Initial production was about 42,000 BOPD.

By 1971, Phillips Petroleum Europe-Africa had been transferred

Ward W. Dunn was awarded Commander in the Royal Order of St. Olaf in Oslo, Norway, in 1976.

from Brussels to London, and I had been promoted to Senior Vice President.

Phase II was designed to bring production to 300,000 BOPD from the Ekofisk field, while returning all produced natural gas to the reservoir. It provided for full development of the Ekofisk field, including the installation of six major platforms, three bridge supports, a one-million-barrel underwater storage tank, reconnection of the original four subsea wells, and the drilling of 38 wells (including eight injections wells). The underwater storage tank, 90 meters (295 feet) tall and 64 meters (210 feet) in diameter, was built of reinforced concrete near Stavanger and towed out to location. This was a unique engineering feat, and the tank originally was to be used to store oil during periods of bad weather to avoid well shutdowns and lack of continuity in production. Later its deck was used for many producing and communication facilities. Drilling operations started on September 3, 1973, and by March 1974, production had increased to about 80,000 BOPD. By 1975, the field was producing at the rate of 300,000 BOPD. The offshore loading system was used until October 1977, when the crude oil pipeline to Teesside became operational.

The latter part of the Greater Ekofisk development, I observed from afar. I was not engaged in the day-to-day operations, having been transferred to the Bartlesville headquarters in 1973 as Assistant Manager for International Exploration. A year later, I became Manager, and later was named Vice President of Exploration and Production for Latin America-Asia. In 1981 I was elected Vice President for Exploration Worldwide.

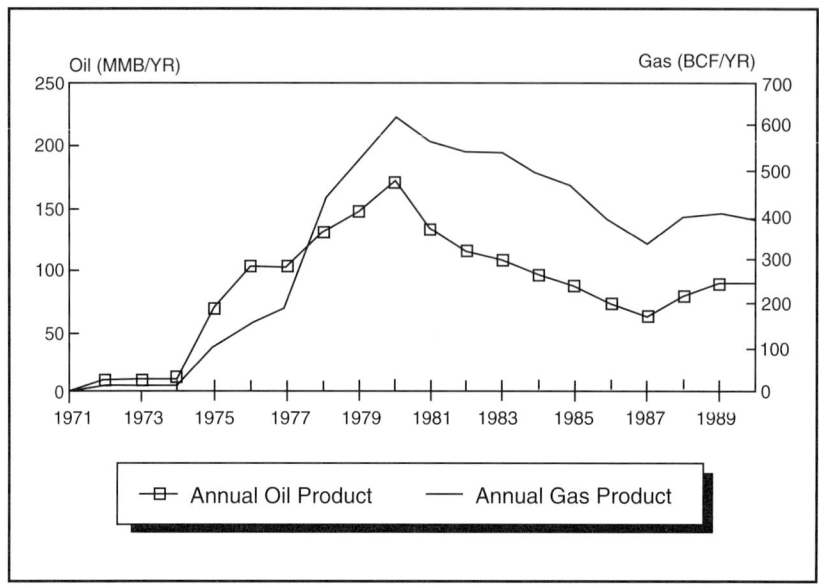

Ekofisk area production, 1971-1990.

In the meantime, Phase III was designed to develop, produce, and process the crude oil and natural gas from the six other fields (Cod, Tor, and West Ekofisk fields, and the subsequent discoveries Abuskjell, Edda and Eldfisk fields)—making a total of seven fields. A 354-kilometer (220-mile), 34-inch crude oil pipeline was laid to Teesside, England, and a 268-mile, 36-inch gas pipeline to Emden, Germany, with attendant shore facilities. When this pipeline and the shore facilities were completed, the system had a capacity of 1,000,000 BOPD and 2.2 billion CFGD. Estimated cost at that time for the overall project was about $2 billion, or 14 billion Norwegian Kroner.

Gas was started down the pipeline to Emden, Germany, on September 17, 1977, and by September 1979 it had reached a level of 1.6 billion CFGD. Ekofisk itself contributes 600 million CFGD of that total. Oil began flowing through the pipeline to Teesside, England, on October 15, 1977 and a year later was averaging 500,000 BOPD, with Ekofisk producing 235,000 BOPD of that total.

Even after oil had been discovered at Ekofisk in 1969, there were many disbelievers. One that I remember was a comment by Sir Eric Drake, Chairman of British Petroleum, on April 15, 1970. He said he did not rate highly the chances of commercial oil finds in the North Sea, and stated, "There won't be a major field there." Six months later, his own company announced a discovery well that "produced sweet oil at the rate of 4,700 BOPD." This was the first major oil strike (Forties field) in the United Kingdom portion of the North Sea. Other companies disagreed with our optimistic forecasts because of the type of reservoir we had found at Ekofisk. Chalk reservoirs had had a history of poor production.

EPILOG

Today, more than 20 years later, the seven fields of the Ekofisk Project are producing at a combined rate of 179,000 BOPD. The Ekofisk Field alone contributes about 66,000 BOPD. The peak occurred in 1980, with production at a rate of 570,000 BOPD for the seven fields. The Ekofisk Field alone was contributing as much as 350,000 BOPD in 1976. Production from the seven fields had declined by 1987 to about 125,000 BOPD. A water flood was initiated with positive results, and I understand the flood will be expanded.

Total production from the seven fields through December 1989 had reached 1.5 billion barrels, of which Ekofisk alone had produced about 930 million barrels. With the water flood, oil production exceeded one billion barrels, a figure we estimated at the time of the initial tests on the discovery well. In addition, the seven fields have produced 4.9 trillion SCF of gas and 141 million barrels of NGL.

Our original cost estimate of about $2 billion has risen to $6.5 billion, which now includes waterflood, correction of subsidence problems, and outlying platforms.

In reflecting now on the discovery of Ekofisk, it is hard to comprehend the amount of risk, energy, engineering skill, and money involved in this pioneering exploration and development project. We were operating at the limit of knowledge at that time, and there were many firsts—such as the longest underwater pipelines, the concrete storage tank, and others. But most importantly, we had the courage to push ahead on that exploration and development project, in the face of tremendous odds and against the advice of many pessimists. The discovery, which was nearly thwarted, transformed Phillips from a domestically oriented U.S. oil company into a truly international competitor, and opened many new and challenging opportunities for its employees. This was as major a turning point for Phillips as it was for Norway. In turn, it also opened many opportunities for Norwegians and for Norwegian companies, and brought many dramatic social and economic changes. It thrust Norway onto the world scene as a major oil producer, spawned Statoil, a state oil company, and gave the Norwegian government a whole new source of significant income. It made Norway self-sufficient in oil—she was no longer dependent on oil imports—and gave Europe a closer, and more politically stable, major source of energy.

Development of Production Sharing Contract

An Indonesian Experience

By Donald F. Todd

"Those goddamn Billings cowboys!" commented one major oil company when they learned that a tiny, underfinanced company had signed Indonesia's first offshore Production Sharing contract.

This contract, signed by Independent Indonesian American Petroleum Company (IIAPCO), gave the rights to explore 57,000 square kilometers (14,000,000 acres) Offshore Northwest Java. IIAPCO subsequently was awarded an adjoining 130,000-square-kilometer (32,000,000-acre) block called Offshore Southeast Sumatra. Together, these areas totaled 187,000 square kilometers (72,000 square miles) —an area comparable to the state of Oklahoma. From the beginning of production in 1971, through the end of 1990, 1,750,000,000 barrels of

Location map of the two contract areas signed by IIAPCO and the Indonesian government.

oil, LPG, and gas equivalent have been produced from these areas. There are more than a billion barrels of proven remaining reserves. Current production is 420,000 barrels of oil and gas equivalent per day.

The contract that IIAPCO signed with the nation of Indonesia became a worldwide model, changing the legal relationship from the colonial concept of concession to one of production sharing—a more realistic partnership between a host nation and a foreign investing company. This change from royalty arrangement to true partnership, allowing Indonesia a voice in management, disturbed the international major oil companies. When IIAPCO's signing became public knowledge, a number of the majors applied strong pressure on the Indonesian oil ministry and on the U.S. Embassy in an attempt to abort our contract.

They didn't succeed. Today, the production-sharing concept is the prevailing type of oil contract used by developing nations.

IIAPCO's success in obtaining this contract and the significant discoveries associated with it heralded the strong entry of independents, large and small, into exploratory efforts around the globe.

This is the story of how I happened to go to Indonesia, what made my associates and me think we had a chance for a contract, how broke we were, how we financed the project, and how it succeeded.

THE BEGINNING

Patrick J. McDonough was an independent oilman from Billings, Montana. Lawrence Barker, Jr., who had recently moved to Denver from Billings, had been trying unsuccessfully to make it big in the northern Rockies since the early 1950s. I was an independent geologist in Billings. I had worked with and been associated with Larry since 1953.

Larry Barker came from a prominent Los Angeles family and had graduated from Yale with honors in mathematics. In 1951, he arrived in Billings to purchase Williston basin mineral interests. In 1953, I sold him on his first geologic lease acquisition. I outlined a 20-square-kilometer (5000-acre) buying area. He proceeded to expand it to 80 square kilometers (20,000 acres). That was the kind of a man he was.

By the late 1950s, he was the most active operator in central Montana, and our company, Tyler Oil, had become one of the largest lease holders in the state. We were wildcatters who had a number of modest discoveries. The scale of our activities, however, exceeded our $2.25-per-barrel-of-oil income. Larry gambled hard, invariably sticking his neck out too far. I gave up my consulting practice in 1955 to join Larry as a junior partner because I thought he was smart and clever enough to run a Texaco-sized oil company. We were to work together for the following 26 years.

In 1978, Larry wanted to take what we had and leverage it into a $200,000,000 public company. Since I had started with nothing, a leveraged business style was not appealing to me. Neither was a public company. I sold Larry my interest in Southern Cross, the company I had formed after leaving IIAPCO.

In 1982, Larry all but leveraged himself into bankruptcy. Since then, he has worked his way to the top once more with a new Indonesian contract and a new public company, Pan-Pacific Petroleum, which has had several recent discoveries in South Sumatra. Larry is now 70 and going strong—the next Armand Hammer. He may yet create his $200,000,000 company!

Pat McDonough, who was always an individualist, received a law degree from the University of Montana. His brother, Jim, financed both their educations by dealing cards and gambling. Pat started a Williston basin completion and workover company. Jim ran the company, while Pat, following another new thought, could generally be found on an airplane headed for wherever. Pat's last great idea before he died was the Northern Tier Pipeline, proposed to carry Alaskan oil across the northern United States into the midwest. It almost worked.

I was the only one without assets, cash flow, or a rich uncle. My industry existence up to the time of the IIAPCO venture had been hand-to-mouth. I had graduated from the University of Michigan in 1950 with a degree in geology. Without a master's degree, however, I was unable even to get a job interview, so I hung out my consulting shingle in southern Kentucky and began looking for oil at 60 meters (200 feet).

I found some oil, but the well went to water after three weeks. (My interest paid enough for me to purchase a quarter of beef, half a hog, and a dozen chickens.) Eighteen months later, I accepted a job with Stanolind in Billings, Montana. Eighteen months after that, in 1952, I resigned to reopen consulting offices in Billings. The time on my own after graduation spoiled me for the corporate world. I was 38 when I first became involved in the Indonesian project, and 41 when the contract was signed.

As an independent, I had needed to have an understanding of lease acquisition and contracts. I needed to stretch exploration funds, to learn to sell a prospect, and to raise funds. A great geologic concept lacking exploration financing is worthless. I had learned to make geologic decisions using low-cost data combined with common sense. Because of limited finances, many of my Montana prospects had been based on photo, surface, and subsurface work. Because I was able to make lease acquisition dollars go farther in the boonies, much of my Montana effort leaned toward frontier exploration. This training proved beneficial, not only during negotiations, but in IIAPCO's exploratory start-up period.

In the fall of 1963, Pat read an *Oil and Gas Journal* article reviewing world production statistics. News of Indonesia intrigued him. Our venture started simply with his saying, "Damn it, Todd, let's go to Indonesia!" I had plugged a favored central Montana prospect the previous day, and I was ready for a change even though I hardly knew where Indonesia was. Montana was (and remains today) a difficult place to make it in the oil industry. In 1963, $2.25 oil and $.06 gas didn't help.

Immediately, I began boning up on Indonesia. I searched the United States Geological Survey library. I went through files at the University of Michigan. A Brigham Young University geology professor, who had spent two years at the University of Jogjakarta, allowed me to copy his notes and articles. Wherever I traveled on business or pleasure I made it a point to search available files.

In the 1920s and '30s, the Dutch had done a great deal of surface mapping in Indonesia and much of it had been published by Van Bemmelen. I sought this and other out-of-print publications through European and domestic rare book dealers. I spent a great deal of time in the eight months prior to my first Jakarta trip accumulating data and trying to define areas of interest.

Pat went to Indonesia in early 1964. I stayed at home, continuing to research the geology, production history, and cultural background of this fascinating area.

Indonesia had a long oil history. North Sumatra was the birthplace of Royal Dutch/Shell as an oil company in 1883. Prior to World War II, 2% of world oil production came from Indonesia. Production was shallow. Many excellent fields were less than 1200 meters deep (4000 feet), and the average depth was only 550 meters (1800 feet). Shell, Stanvac, and Caltex were responsible for nearly all of Indonesia's production in the pre- and post-war years.

The 1949 revolution in Indonesia ended 300 years of Dutch rule. President Sukarno, a strong, adventuristic leader, ruled under what he termed "guided democracy." The PKI party, affiliated with Chinese communists, became a more and more dominant influence on Sukarno's PNI party and on Indonesia.

The political climate was becoming increasingly difficult for foreign businesses. Few western companies bothered to even look at Indonesia. Oil, rubber, and a few other industries with pre-war ties endeavored to continue under adverse conditions. In late 1965, Shell gave up and sold out.

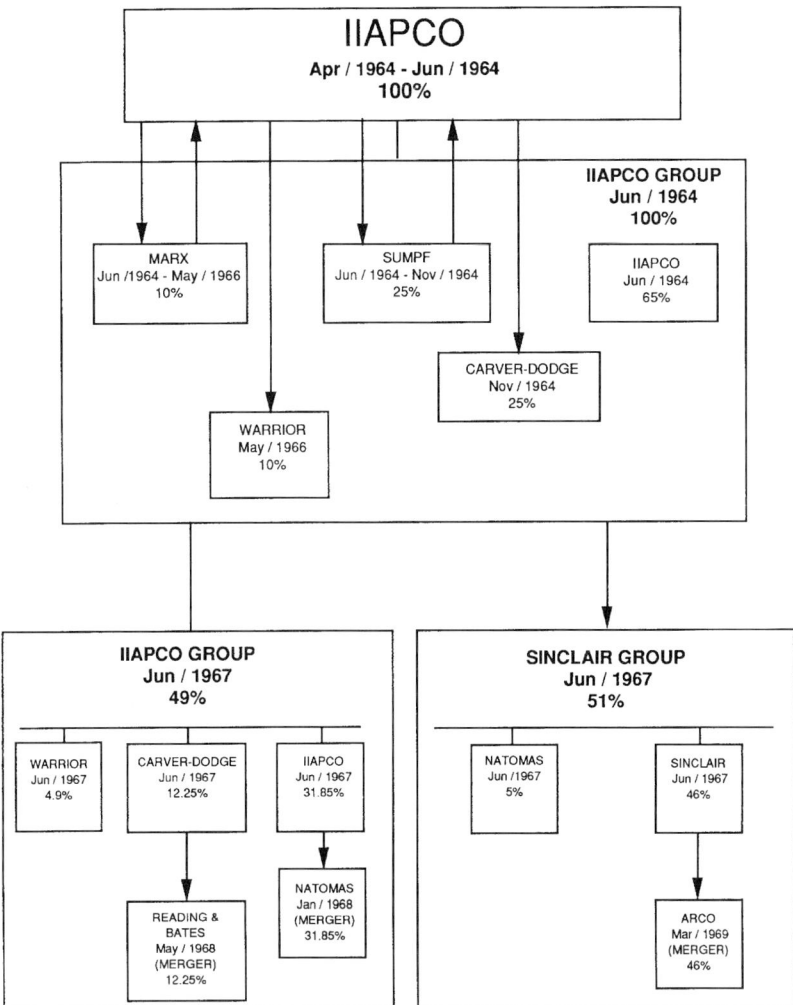

IIAPCO's Indonesian rights and ownership through June 1969, according to the Offshore Northwest Java Production Sharing Contract (signed August 19, 1966).

Against this unpromising backdrop, Pat, Larry, and I formed our company in the spring of 1964 and sought an Indonesian oil exploration contract. Larry was president, I was executive vice president, and Pat secretary-treasurer. Pat picked the name Independent Indonesian American Oil Company, known as IIAPCO for short. [In 1968, the Natomas Company, a shipping and real estate concern, acquired IIAP-

CO and later wanted to drop the name, but they were told it belonged in Indonesia. Natomas's successor now uses Maxus as an operating name in Indonesia.]

In June 1967, Sinclair Oil Company joined IIAPCO's Indonesian venture. To formally introduce Sinclair to the Indonesians, we met with the five men who then constituted Suharto's Presidium Cabinet. When one of these ministers was introduced to me, he said, "Of course, I have heard much of IIAPCO." When he was introduced to Bob Elston, president of Sinclair International, he said "Yes, I believe I've heard of Sinclair, too. Don't you can sardines?" IIAPCO had gained its competitive advantage.

The frequent rumors of bribery of Indonesian government officials had results we found both embarrassing and amusing. So sure were some detractors that Major General Dr. Ibnu Sutowo, head of Indonesian oil, had his hand in our pocket that IIAPCO was frequently referred to as "Independent Ibnu American Oil Company." There was no way to convince Ibnu's accusers that he was honest. Indeed, Ibnu was the most trustworthy person I had ever had occasion to deal with in the oil business. There are still those who believe IIAPCO could not have obtained a contract without a bribe.

When IIAPCO was formed, it had exactly three shares of stock. McDonough had a share, Barker had a share, and I had one. I owed Larry $18,000 for previous advances. I knew no other way to pay him back, so I offered to work it out, contributing his share of my *per diem* charge for this Indonesian venture. In 1966, on bringing home the signed contract, Larry commented, "I never really thought you would get a contract. Now I believe I could sell my share for $50,000."

We decided that political unrest, lack of a good road network, a tropical jungle environment, and a dense population precluded most onshore areas from consideration. We were also aware that initial seismic exploration would be faster and cheaper offshore. This was important to IIAPCO, a company short on finances. I had selected four areas of interest: the eastern Kalimantan offshore, the southern Kalimantan onshore portion of the Barito basin, the offshore area around Madura Island, and offshore northwest Java. We had no detailed information on any of these areas, but we hoped to obtain geologic data in Indonesia that would help refine our final area selection.

None of us had any previous involvement in foreign exploration. Our industry efforts had been focused in the northern Rockies.

EARLY CONTRACT EFFORTS

Pat and I headed for Indonesia in June 1964 with little more than plane fare in our pockets. On the way, I stopped in Los Angeles to visit with Bob Sumpf, a geologist and sometime investor who had done business with Larry and me. Bob and an associate, intrigued by our Indonesian project, agreed to take a 25% interest by paying 50% of costs up to a total $25,000 group expenditure. Thereafter, IIAPCO would pay its share. Larry promoted Louis Marx (later a principal in the formation of Pan-Ocean) for a 10% interest on a similar basis. Our out-of-pocket costs, including a *per diem* for me while in Indonesia, were now 70% covered.

We were on our way! Pat and I never considered the possibility that we might not succeed. We island-hopped across the Pacific and Southeast Asia: Hawaii, Wake, Guam, Manila, Saigon, Singapore, and finally Jakarta (this was before 747s). Our American contact in Indonesia on this trip and the next was Bill Palmer, a representative of the American motion picture industry and a friend of President Sukarno. However, the PKI, the Indonesian communist party, thought Bill to be a CIA representative. "GO HOME BILL" slogans were everywhere.

In 1964, there were three Indonesian state oil companies. Permina, controlled by the army, operated in North Sumatra. Its president-director was General Ibnu Sutowo, a strong pro-West entrepreneur. Permina had previously entered into two simple prototype production-sharing agreements, one with Refining Associates, owned by Harold Hutton, and the other with Asamera, run by Tom Brook. In late 1965, Permina bought out Shell's Indonesian interests.

Pertamin, controlled by Sukarno's PNI party, was the state enterprise responsible for domestic marketing. It oversaw Caltex and Stanvac operations in Central and South Sumatra as well as Pan-American's new work program in south Sumatra. Permigan, the third state enterprise, was the PKI communist oil company responsible for Java and Madura.

We searched every Jakarta bookstore, but were unable to find any more geologic information. We decided to try to obtain a contract despite the fact that our geologic data bank left much to be desired. The choice to try for offshore northwest Java (though IIAPCO would then need to negotiate with the politically most leftwing of the three state oil enterprises), boiled down to very simple political, logistical, and geologic reasoning. What little was published on western Java geology indicated that the Sunda shield—the effective basement—would be encountered a few miles offshore.

IIAPCO did not concur with this thinking. We felt the lack of geologic consideration given the area by others would be advantageous to us in negotiations. Little interest had been expressed in Indonesia's offshore. Since this offshore area was physically and politically remote from the mainland, we believed that this would lessen our political problems.

The areas of interest in Kalimantan and Madura had many logistical disadvantages, particularly to novices (which we certainly were). Jakarta, on the coast, adjoining our selected area of interest, is the largest city in Indonesia. It has the best seaport and, at the time, it had the only international airport. The Java Sea is the largest shallow shelf sea in the world, with an average depth of 37 meters (120 feet). It is essentially devoid of bad storms. Operations in this area would be easiest.

Data on offshore northwest Java were more limited than on our other areas of interest. However, the narrow strip between the volcanic chain and the coast had numerous recorded seeps. The first oil exploration effort in Indonesia, three 30-meter (100-foot) tests drilled in 1871, was onshore near Ceribon. Although we had been informed by an Indonesian geologist that the sediments along this coastal strip contained a great deal of volcanic material, which might result in poor reservoir-quality rocks, we decided that real data were too insufficient to draw any conclusion. We were unable to obtain an electric log anywhere in Indonesia. Charlie Doh of Schlumberger later slipped me a couple of logs for Sumatra out the back door. Only after our contract was signed and ratified were we able to obtain a few electric logs from the Indonesians.

Faced with this lack of data, we went to the Indonesian navy and obtained the Dutch bathymetry of IIAPCO's offshore areas of interest. By contouring the water depth of the Java Sea, it was easy to define an area of interest for offshore northwest Java bounded on the west by the Thousand Islands, on the north by the Karimundjawa Islands, and on the east by a sharp coastal change near Semarang. This was a simple task for one who relied on surface, photo, and drainage work for the creation of many Montana prospects.

To my mind, it was evident that a basin existed off the northwest coast of Java. It was also apparent that if the water level surrounding South Sumatra were raised 15–30 meters (50–100 feet), nearly every oil field would be submerged. If the water level off Java were lowered 30 meters (100 feet), there would be an extensive plain not unlike Sumatra. During the Pleistocene, the entire floor of the Java Sea was a land bridge connecting Java with the Asian mainland. With a little imagination and wishful thinking, it was easy to connect the Sumatran productive trends across IIAPCO's offshore area of interest, and to extend this trend around the archipelago to the productive areas then known in eastern Kalimantan.

On our first trip, we met with Third Deputy Premier Chaerul Saleh, one of President Sukarno's top men. We were favorably received and given approval to negotiate a Contract of Work with the Ministry of Basic Mining, Oil and Natural Gas, and with Permigan.

We borrowed two typists from USAID, and with a copy of Pan-American's 60-page Contract of Work as a guide, we endeavored to draw up our own. Pat thought we should have the contract finished and approved in a week.

While the Indonesians reviewed the first draft, we decided to take a few days to see Java. Near Ceribon, near IIAPCO's proposed area and on the flank of a large volcano, we happened upon two old permanent derricks in the Bongas field. We found someone to open these wells for us. One well drilled by Shell in 1941 was capable of flowing 40 barrels of oil per hour (BOPH) of 40° API oil. The other, drilled in 1939, was capable of flowing 25 BOPH. Pat and I were now sure we had selected the right area.

On June 30, 1964, an article in the *Jakarta Daily Mail* stated that the government was having discussions with four oil companies, including IIAPCO, and that Indonesia was considering reopening negotiations for cooperative exploratory efforts. Subsequent meetings slowed as officials presumably awaited reaction to the newspaper article. On July 8, the same paper carried an interview with the Information Minister. He warned Indonesians that "foreign-owned oil companies operating in this country were the claws of imperialism and capitalism for which the people should be most careful and vigilant." The game had commenced.

Each day we made calls on a number of offices. We met and made friends with many Indonesians in the government and state oil enterprises. No one, however, would tell us what was going on with our contract. Finally Minister Saleh's lawyer told us everything seemed OK except that the government expected a small signature bonus.

Communications from Jakarta were difficult at best. As Indonesia was on bad terms with nearby Malaysia, we couldn't go to Singapore to call the U.S. This necessitated a 10,000-kilometer (6000-mile) round trip to Bangkok in early July, just to make a phone call. Even then, radio phone connections were bad. Pat was frustrated and decided to return to the States.

When I got back to Jakarta, I tried to speed up negotiations. Saleh's lawyer informed me that as IIAPCO was seeking a 30-year contract, a three- or four-day wait now and then shouldn't hurt a bit! The true concept of patience was first introduced to me by the Indonesians. I was glad Pat had gone home. These gentle people didn't understand Pat's hard-drinking, big-hearted, but impatient and brusque nature.

Finally, I was told we would have to agree on a bonus before any contract would be approved. I returned to the U.S. at the end of July to meet with my associates in an endeavor to meet the Indonesian demands. We began working on a revised contract.

It turned out that Permigan wanted a $3,000,000 bonus. I sent a message back that we had anticipated a request for a token bonus, not one that would finance the country. We gave them the short course on how independents work. Although independents were always short of capital, we explained, they often reached the finish line before the major oil companies!

In mid-September, we sent a revised contract draft offering a $300,000 bonus. The Indonesians found the offer offensive but indicated they were willing to negotiate. In December 1964, we received a cable inviting IIAPCO back to finalize the contract. All we had to do was come to a compromise on the bonus.

Bob Sumpf, a confirmed anti-communist, had been reading too much bad Indonesian press. He decided he wanted no part of doing business with a pro-communist government and declined to support a second trip. Dave Dodge, a geologist and acquaintance of mine, quickly committed Carver-Dodge Oil Company to pick up Sumpf's 25%. Dave and his partner Doug Carver became IIAPCO's strongest supporters.

In January 1965, Sukarno announced that Indonesia was withdrawing from the United Nations. There was evidence that Sukarno had formed a working alliance with Red China to force the Anglo-American presence out of the area. Sukarno had recently told the U.S., "To Hell with your aid." In spite of these confusing and disturbing actions, I decided to return as planned.

Still broke, but with 70% of our costs covered, IIAPCO returned to Jakarta in mid-February 1965. En route, I learned that the U.S. Information Service (USIS) in Jakarta had been taken over and the American flag destroyed. This was followed by a takeover of another USIS office in Medan, Sumatra.

The atmosphere in Jakarta was one of some desperation. I arrived in Jakarta only to be invited to a cocktail party for departing U.N. personnel. It was difficult to obtain any meetings with Permigan and Indonesian government representatives. However, I was still optimistic that I would have a contract signed within two weeks.

Inflation was rampant. The oil companies could get their rupiah (Rp) only at the official rate of 45 Rp to the dollar, while the black market rate in early February reached 15,000 Rp to the dollar. It was necessary to carry a briefcase full of rupiah just to go to a Chinese restaurant for dinner. Gasoline purchased at the black market rate of exchange was about one cent per liter. You could purchase 250 liters of gasoline for the cost of one whiskey sour at the Hotel Indonesia.

When I met with Saleh's lawyer to discuss the bonus, he indicated that Indonesia would not accept our $300,000 offer, and I told them IIAPCO wouldn't accept their $3,000,000 proposal. I said that our top bonus offer would be $1,000,000 in exchange for certain changes in the contract. The Indonesian economy was so bad that I hoped they might accept this offer, even though I knew that Pan-Am had paid a $5,000,000 signature bonus when they signed in 1962.

Meanwhile conditions were rapidly deteriorating into a serio-comic version of *The Year of Living Dangerously*. A few days later, Bill Palmer's dog was kidnapped and held for ransom. Americans were being verbally insulted in the streets. On February 26, 21 newspapers and magazines were shut down. Everywhere, the PKI was flexing its muscles. Dramatic changes for the worse had occurred since our previous summer trip.

The Indonesians are intelligent, delightful, and friendly people with a keen sense of humor, and their family relationships are more like our own than elsewhere in Asia. I continued to call on Basic Mining and Permigan officials, and began to visit them at home and meet their families. These close relationships were to prove most beneficial after the aborted September 30, 1965 coup.

On March 1, 1965, First Deputy Premier Subandrio asserted that Indonesia's policy of confrontation would continue. Communist-led demonstrators took over the U.S. rubber plantations and factories. Newspapers were calling for the arrest of Bill Palmer, claiming he was a CIA agent. There was a call for all American newsmen to leave. Five hundred demonstrators entered U.S. Ambassador Jones's residence. I had returned to Indonesia at the government's invitation, yet after a month, I had not had a formal meeting with them. I had, however, learned patience and I remained determined to go home with a signed contract.

Political turmoil made me a hotel prisoner. To pass the time, I requested permission to review Permigan's geologic files on the adjoining onshore area. I was told that this would be permitted only after a contract was signed.

Indonesian offices closed at 2:00 p.m. After that, foreigners staying at the nearly empty Hotel Indonesia congregated at the swimming pool to exchange tales. I met a displaced Czechoslovakian-U.S. entrepreneur who dealt in everything from human hair for wigs, to quinine, to drill bits. "Politics is lousy," he commented, "but business is goot." I reflected on a statement I had read when first considering this Indonesian venture: "better to have an area of good geology and sorry politics than good politics and sorry geology."

Sutan Assin, a petroleum engineer and the one Permigan director who was not pro-communist, had become a good friend. He discussed Indonesia's profit-sharing concept whereby the foreign company furnished the money and Indonesian management spent it. Ir. Wijarso of Basic Mining, and Hanafin, Saleh's lawyer, talked "profit-sharing" and told me all "Contractor" deals were a thing of the past.

IIAPCO's refusal to pay a large bonus combined with the current political unrest had changed the game. I informed them that they needed the expertise of people like me to help bring Indonesians into the oil industry. They replied that they would like to hire me if my price wasn't too steep. On March 10, 1965, I cabled home: "OLD DEAL HAS BEEN THROWN OUT. THEY NOW WANT IMPOSSIBLE PROFIT SHARING. AM TRYING TO PULL IT OUT BUT DOES NOT LOOK GOOD."

That was the understatement of my life. Everything had blown up. Shell was given 48 hours to appoint an Indonesian director. Association with Americans was becoming difficult for Indonesians. A PKI rebellion in Surabaya resulted in the shooting of 17 Indonesian navy officers. I decided it was time to pack my bags and go home, my dream of an IIAPCO contract shattered. We could only be thankful we hadn't spent $25,000,000, as Pan-Am had on their contract.

I left Indonesia on March 13, 1965. The plane was loaded with Americans and other foreigners. Even Bill Palmer was aboard. Indonesia had been his life. The following day, his motion picture offices and his Bogor mountain retreat, where I had spent many pleasant weekends, along with all his personal property, were confiscated. Sukarno announced that the government was taking over management of Shell, Stanvac, and Caltex. Electricity and postal service to many Americans was disrupted.

I returned to Billings, discouraged. Our Indonesian venture was kaput and I thought I would never see that part of the world again. To tackle this project had necessitated a virtual shutdown of my one-man office. I had to start all over again in Montana.

Like everyone else, my partners and I followed the deteriorating Indonesian political climate via the news, and we maintained contact with the U.S. Embassy in Jakarta. None of us was prepared for the culmination and failure of the PKI power struggle and attempted coup of September 30, 1965. Six generals, the policy-making core of the Indonesian Army, all with friendly relations with Westerners, were ambushed and killed by the PKI at Air Force headquarters. Only quick action by General Suharto aborted the coup. Months of indiscriminate bloodshed followed. It was difficult to know what was happening.

SUCCESSFUL NEGOTIATIONS

In time, the stormy political weather in Indonesia began to clear. In February 1966, Dave Dodge encouraged me to consider going back to Jakarta for another try. Through the U.S. Embassy, we learned that Major General Doctor Ibnu Sutowo, who had long run Permina, was now also Minister of Mining, Oil and Natural Gas and Director General of Oil and Gas Affairs. We corresponded with General Ibnu through the embassy. Part of our March letter stated:

> IIAPCO is neither owned nor controlled by a large American oil company. It is composed of a group of small independent companies who have available sufficient resources to do this major work. We ask for the opportunity to explore a now unknown area for oil and gas, to spend the agreed-upon sums for exploration and drilling, and, in the event of the discovery of production, to be allowed to recover our costs and make a reasonable profit commensurate with the risks which we are prepared to take.
>
> We are completely in sympathy with the desire of the Republic of Indonesia to control and develop its natural resources and to prepare its own nationals to manage this important work. If this can be accomplished within the framework of an agreement which will protect and benefit both parties to the agreement, IIAPCO would welcome the opportunity to present our proposal again.

Paul McCusker, U.S. Embassy Counselor for Economic Affairs, met with General Ibnu on our behalf. Ibnu indicated that he would welcome our visit, but that any proposal must follow the lines of the contracts Permina had signed with Refining Associates and Asamera. He indicated the Pan-Am Contract of Work we had been following would not be acceptable, and he made it clear that negotiations would not be along lines of "profit" sharing, but rather of "production" sharing.

In May 1966, I began to shut down my Billings business for another try at Indonesia. By this time, Louis Marx, who had maintained his 10% interest through our two contract endeavors, decided not to support IIAPCO's third effort. I owed my bank $6,000, money I had borrowed so that my family could exist while I was in Indonesia. Pat refused to pay me his share of my *per diem*. When Marx dropped out, IIAPCO had a chance to resell this interest and perhaps get enough up front to pay off my banker, who felt I shouldn't be heading for Indonesia without first satisfying the bank.

Down in Denver, geologists Will and Joe Obering ran a family company called Warrior Oil. I had never been able to sell them anything, but they always had time to listen. For some time, I had been presenting deals to them without any expectation of their buying. It was just a good way to get the bugs out of my pitch.

When I got to Denver, Will was out of town, so I gave Joe my presentation. My proposal was the same: 10% interest for 20% of costs except that we asked $10,000 up front for previous trips. To my surprise and delight Joe committed! My debt was paid off and IIAPCO had $4,000 in the bank! When Will returned, he bitched loudly about Joe's commitment, but he really didn't want out. Alone of the original group, Warrior Oil today retains a direct interest in the Indonesian IIAPCO operations.

On June 7, 1966, about the time I was scheduled to leave for Jakarta, IIAPCO received a cable from the U.S. Embassy: "IN LIGHT RECENT REPORTS CONCERNING STATUS GENERAL IBNU SUTOWO/EMBASSY SUGGESTS TODD DEFER TRIP PENDING FURTHER DEVELOPMENTS." It was a disquieting communication, but I set out for Indonesia anyway, after waiting a week and conferring with my associates.

Ibnu's job was a powerful one, ultimately connected with money. Labor unions and student groups were pushing for Ibnu's removal on grounds of corruption and mismanagement, but there was a growing Indonesian realization that foreign investment was necessary.

While I was en route, I learned that Third Deputy Premier Chaerul Saleh, with whom I had previously been dealing, was in jail. Former First Deputy Premier Subandrio and 14 others had been taken into custody. More and more power was being given General Suharto, and Sukarno was fast becoming a figurehead. It was rumored that Ibnu was under house arrest.

To my relief, on arrival I learned Ibnu was not under arrest. Anondo, Sutan Assin, and Wijarso had also survived the shuffle. All remained under Ibnu either in the Ministry, the Directorate, or with Permina. IIAPCO was fortunate that many of the Indonesians we had met with on previous trips remained in positions of importance. They considered offshore northwest Java as an area reserved for IIAPCO. Wijarso had told General Ibnu about IIAPCO's previous efforts and how Pat and I turned out a contract draft in two weeks. Permigan, the company IIAPCO previously had worked with, had been liquidated because of its PKI affiliations. The "GO HOME YANKEE" signs were gone now, replaced by anti-communist slogans.

IIAPCO's limited finances precluded my taking a lawyer with me. As an independent, I was accustomed to reading contracts. Bob

Fowler, a lawyer associated with Carver-Dodge, was to be my stateside backup. But with no reliable telephone contact, and letters taking as long as three weeks and cables taking eight days to arrive, I was essentially on my own. Fortunately, Indonesia's confrontation with Malaysia had been called off and it was now possible to go to Singapore to communicate. This was only a 2000-kilometer (1200-mile) round trip.

After making initial protocol calls, I expected the usual long wait for a formal appointment with General Ibnu. I hoped an appointment could be arranged for the following week. Friday was a Muslim holiday, so I decided to go to the beach for the weekend. When I returned Sunday night, I was chagrined to learn an appointment with General Ibnu had been scheduled for the previous day. I thought I had blown it. Early on Monday I went to Permina's office to apologize and was told Ibnu had rescheduled the meeting to start in a few minutes.

In our first meeting, July 4, 1966, I found Ibnu cold but efficient, and only much later was I exposed to his radiant smile and personal warmth. He laid out the ground rules for his Production Sharing concept. He indicated that the split would be 65/35, except if we desired to consider a large bonus. He asked if I could accept discussions within the framework of his concept. I didn't feel I had a choice.

He asked me how long I planned to stay, and I informed him I would stay until I had a signed contract or he threw me out. Ibnu liked that. Most foreigners wanted to come one day and go home the next, he said. When he asked if I was ready to start and I replied that I was, he looked at his watch and said "How about in 30 minutes?" What a change from the long waiting periods of previous trips! Then Ibnu indicated that he had received letters from several companies expressing interest in Indonesia. I knew Japex, Tenneco, Union, and Zapata were interested. The race was on!

I soon met with Dr. Sanger and Mr. Kwa (Tirto Utomo), the Permina lawyers. We discussed contracts, and they gave me a draft Production Sharing Contract drawn by Hank Brandon of Union Oil Company a couple of years earlier. Though I felt the draft needed some modifications to satisfy IIAPCO, it was well thought out. (Later when I met Hank, he told me how hard he had worked on preparing it, only to fail to get Union's approval to finalize a contract.) Ibnu had signed two earlier and much simpler versions of his Production Sharing concept with Asamera and Refining Associates.

Sutan Assin now held two jobs. As head of Permina Unit III, he was specifically charged with Java and Madura, an area previously handled by Permigan. He was also head of exploration and production for the entire country. He told me he thought IIAPCO would get a contract this time.

Indonesia was still in a state of confusion and unrest. No contracts had been signed with foreign investors in any industry since the aborted coup nine months earlier. Although the U.S. Embassy was generally pessimistic about IIAPCO's chances, they remained cordial and helpful. Because normal communication channels were difficult and time-consuming, embassy staff agreed to open the diplomatic pouch to us to send contract drafts to the U.S.

The embassy made me aware of Ibnu's political insecurity. I was assured by Permina and the ministry that if a contract could be signed before a cabinet change, it would be honored. If Ibnu were to fall, it might be a year before the government would again be ready to negotiate. Knowing several draft contracts would be necessary before a signing would be possible, the Permina negotiating team and I felt we had to work fast to finalize the contract before more political change could occur. A code was set up so my stateside associates could quickly suggest changes, additions, and approvals. I did not want to waste time traveling to Singapore.

The Permina negotiating team was headed by Dr. Sanger. Working with him were Mr. Kwa, Mr. Siregar, an accountant, and Dr. Sakidjan, an economist. Kwa, a brilliant Indonesian of Chinese extraction, was the spokesman and perhaps the most important member of the team. He was the man with whom I spent the most time and he had a wonderful sense of humor. Indonesia's subsequent ethnic unrest mandated that all Indonesians of Chinese origin assume Indonesian names. So Mr. Kwa became Tirto Utomo. "Tirto" is now a very successful Jakarta businessman.

Demonstrations were common. Gun-toting soldiers, armored cars, and barbed-wire barricades were everywhere, and uneasiness and unrest continued. Jakarta, a city of 4.5 million, had but one traffic light. It was one huge traffic jam with people, dogs, *betjas* (bicycle-taxis), cars, and trucks all in the same place—in front of my taxi.

In our meetings, I reviewed previous contract proposals. After spending time with Mr. Kwa discussing changes and additions, I met with the full Permina team. The fact that I was alone and negotiating with four men from Permina actually turned out to be an advantage. Permina agreed to furnish me an office, a secretary, and a "legal helper."

We commenced the first draft at 7:30 a.m. on July 8, 1966. On this first day we completed only four pages. My Permina secretary spoke and typed English well. All negotiations were conducted in English. The contract was being written in English. I had discovered that the average educated Indonesian spoke not only Bahasa Indonesian and a few dialects, but generally Dutch and English, too. This was most fortunate for me. I spoke only English.

My "legal helper" turned out to be T.N. (Roger) Machmud, who was to become IIAPCO's first paid employee. Roger was a lawyer who spoke English well, and was familiar with the Permina and government structure. Contract drafting proceeded at a rapid pace, partly because of Machmud's close relationship with Permina. The Indonesians were amazed to learn I was there not only to negotiate but to approve and sign.

Demonstrations and parades organized to undermine President Sukarno and enhance Suharto's political position continued. Although Sukarno had borne a great deal of responsibility for the coup and the subsequent turmoil, he remained Indonesia's George Washington and a hero to the masses.

Although I did not formally meet General Ibnu during negotiations, he always knew what we were doing. Tenneco was now pressuring the U.S. Embassy for more information as to IIAPCO's progress. They particularly wanted to know if I intended to return to the U.S. for our "legal department's" consideration. Daily letters were written home or cables sent to Larry indicating progress, changes, and requested changes, along with current rumors. I was learning how Permina wanted to be treated as a partner. It was apparent that it mattered more to them who I was rather than whom I represented. Negotiations were very personal and relaxed.

On July 10, I joined a group of 40 foreign embassy people, businessmen, and their families on a Sunday excursion to the Thousand Islands on the western edge of our proposed contract area. The army furnished a hydrofoil for this charity event. As someone had neglected to remove the barnacles from the craft, we were unable to foil. After a late arrival at our destination, a lovely day of reef exploration and a late departure we proceeded to run out of fuel about 24 kilometers (15 miles) out of Jakarta. Our anchor wouldn't reach bottom, and we had no radio and no signal lights. Four of the crew went for rescue in a rubber raft. A landing craft with some barrels of fuel returned about 4:00 a.m. and the crew proceeded to carry fuel aboard by pail. I arrived at my hotel at 6:45 a.m. and made my 7:30 Permina appointment. This was my first on-site look at Offshore Northwest Java!

Talks between the Indonesian government, the International Monetary Fund (IMF), and the World Bank were in progress, and expectations were high that substantial grants and loans would be seen before year's end. Indonesia and Permina were particularly anxious to display a newly favorable attitude toward foreign investment, so IIAPCO could not have picked a better time to be there. Although unrest continued, the waters were clearing rapidly, and our competition had not yet arrived.

IIAPCO's first draft was submitted July 12. It was sent by U.S. Embassy pouch to Denver with a plea to my associates to keep their changes to a minimum, to make them simple, and to trust my judgment. I was particularly interested in stateside advice regarding U.S. tax ramifications of our draft contract. I expected to receive Permina's comments by July 16th and to reconvene negotiations on the 22nd.

By mid-July, Sukarno's authority to issue decrees was rescinded, the Malaysian confrontation was officially ended, and an application for re-admission to the United Nations was being made. Sukarno was stripped of his lifetime presidency. The PKI, which had been the third largest communist party in the world during my 1964 and 1965 trips, was now virtually wiped out. A country that had been a Chinese satellite was now an outspoken foe.

IIAPCO was extremely fortunate that groundwork had been laid during a period when Western business wanted little to do with Indonesia. I returned at a time when Permina and Indonesia needed a Western guinea pig on which to try out their new business desires and intentions. IIAPCO's luck should be attributed to a little foresight, a bit of naivete, the optimism of youth, and the encouragement of my associates. Even the embassy was surprised at IIAPCO's rapid progress.

Before negotiations reconvened, Machmud resigned from Permina after a run-in with one of Ibnu's top associates. IIAPCO retained Machmud to help with contract negotiations, after receiving assurances from Kwa that it would not pose a problem.

To head off a new Permina signature bonus request, we suggested instead that IIAPCO give bonuses to postpone relinquishments of our area. In this way, our up-front cash requirements were delayed. We reworded many "colonialist" clauses objectionable to Permina. They said the same thing, but more aptly satisfied the spirit of nationalism and partnership desired by Indonesia. I was informed that IIAPCO's chances were good for several reasons: Permina liked me. They felt IIAPCO had a great deal of courage to attempt exploration in a new area. They needed publicity and had an urgent desire to get things moving.

My stateside associates were busy obtaining bank references to establish partner credibility. IIAPCO still had only three shareholders: Barker, McDonough, and myself. Carver-Dodge and Warrior's interests would result in a direct contract interest if we succeeded. As we wanted to be viewed as one entity, we began to refer to our venture as the IIAPCO Group. It was so helpful in the negotiating phase that we continued as the IIAPCO Group well into the later exploration and production phases.

By now the general opinion in Indonesia was that Ibnu would relinquish his position as minister, but remain as head of Permina and continue to be "Mr. Oil." We went on with negotiations knowing that a new cabinet would be formed, and we assumed that a man of intellect and with an understanding of world business would honor a signed contract.

Events were now moving so fast that often I couldn't wait for Denver's comments and opinions. Fortunately, my associates realized this and allowed me to act on my own. I was not sure why things were moving so fast, other than because Permina hoped to finalize before there was any change in the status quo.

Permina was the top-paying Indonesian state enterprise. Yet Machmud, a smart young lawyer, was making only 1,400 Rp per month (U.S. $13.33) when he resigned. It cost him 3,000 Rp per month just to run his household. In other Indonesian state enterprises, a college-educated professional might earn 300-1000 Rp per month. Even "top-paying" jobs required moonlighting to supplement income.

Denver sent me certified copies of IIAPCO's incorporation papers along with a fancy document appointing me agent and attorney-in-fact. We were getting our act together so that I could sign when the time came.

On days when Machmud and I met with Permina, meetings started at 7:30 a.m. and generally continued until their offices closed at 2:00 p.m. After we discussed progress and strategy, Machmud would have more meetings at Kwa's home while I wrote to my partners or sent cables to Denver. On other working days, we met in my hotel room, which was converted into a daytime office, where Roger, our secretary, and I worked on the contract. I tried to get away from Jakarta after Permina's Saturday closing. Indonesia is a beautiful country to enjoy—rich in art, culture, and history.

In far-off Denver, Bob Fowler was busy advising us on contract wording relating to U.S. taxes, depletion, and title. [One Fowler cable commented, "YOU PRETTY GOOD LAWYER."] With the time delay in mail and cable correspondence and with no viable telephone communication, Roger and I had to think cautiously and carefully before adding any new wording. We needed a contract workable from IIAPCO's side as well as Permina's. It turned out that the red tape involved in sending documents through the embassy pouch slowed rather than speeded our communication process. We ended up sending drafts through both the pouch and the mails. Often KLM posted our mail in Amsterdam.

Many of the late contract changes were for public consumption in this proud, emerging nation. We replaced "IIAPCO shall keep the books" with "Permina shall be responsible for keeping the books...however, Permina shall delegate to IIAPCO its obligation to keep books."

"IIAPCO shall...and Permina shall approve" was included frequently for this reason. I had learned to understand Permina's and Indonesia's needs, and the negotiating team was learning to understand IIAPCO's needs. After Indonesia pushed out the Dutch in 1949, there remained a strong anti-colonialist attitude as Indonesian nationalism grew. This is reflected in the Production Sharing Contract vs. the Concession Contract.

It was to our advantage that IIAPCO was a little company. I had come to realize that Tom Brook was able to obtain a contract because he was the man who could decide for Asamera. The majors generally sent one man to scout, another to talk, and eventually they would send along an intimidating battery of negotiators. Even then, when changes were proposed, they had to return to the U.S. for advice.

IIAPCO's second draft was submitted July 20, ahead of schedule. Indications were that we could have a final version ready for signing by the end of July. Instead of appointments being postponed, as had happened on earlier trips, they were now being advanced by Permina, which was trying even harder than IIAPCO to finalize a contract. We were told IIAPCO was the most progressive oil company Permina had dealt with in years.

In the meantime, the Japanese arrived. They were slow, picky negotiators and Permina wanted to finalize IIAPCO's contract in order to hold it over the head of Japex and to show the rest of the world as well.

We were working on the third draft on July 25, 1966, when the new cabinet was announced. The new Minister of Mining was Ir. Slamet Bratanata. Under his ministry came General Ibnu as Director General for Oil and Gas. Ibnu also remained as President Director of Permina. Bratanata was a civil engineer holdover from Saleh's Basic Mining and Industry. He had never been involved in oil and gas work. Although we were told he was a young opportunist, everyone felt he would have a difficult time saying no to this "progressive" contract. The consensus was that Bratanata was put in to remove some of the union pressure from Ibnu, but that Ibnu's real power would remain. Bratanata's appointment was essentially political—or so we thought.

The Indonesian government was anxious to sign a Production Sharing Contract with an oil company. I was aware that those companies following IIAPCO would find it increasingly difficult and expensive to finalize contracts, because terms would be stiffer. I became so confident of an early signing that I was busy making plans for how to best operate in Indonesia.

We remained concerned about labor unions. They had broken Shell Indonesia's back, and continued to give problems to Stanvac and Caltex. Now they were demanding that General Ibnu increase Permina wages. Ibnu retorted, "go to China." (In a 1970 interview, General Ibnu was asked how he had escaped the influence of communism. He answered, "From the beginning, it was very personal. I thought it a very good theory. However, it will never be possible to implement, simply because it is not based on human beings as they are.... Unless we change human nature, there will never be a chance to have the ideas in communism realized, as it goes against every human instinct that we all have, regardless of where we are, what color of skin we have, or what education we have.")

IIAPCO's first-year expenditure requirement was $300,000. I had some difficulty retaining this figure because Japex was offering to spend a minimum of $750,000 in the same period, but I succeeded in convincing Permina that independents get twice as many miles for their dollars. I wanted to keep initial expenditures low to cover our backsides while IIAPCO found the large-scale financing needed to adequately evaluate an area of 57,000 square kilometers (14,000,000 acres).

A cable from Barker stated, "THE IIAPCO GROUP DESIRES COPY COMPLETE FINAL DRAFT FOR APPROVAL PRIOR TO SIGNING BY YOU AS AGENT AND ATTORNEY-IN-FACT." Though I appreciated their nervousness, I doubted I would be afforded the luxury of the time required to send them a copy. Also, I was running short of money for my living expenses in Indonesia. Transfers were difficult, if not impossible. It took quite a while to find an Indonesian willing to take a Billings check and give me rupiah!

When I visited friends, I took along paperbacks for gifts, mostly fiction, because light reading was difficult to come by in Jakarta. I also brought presents such as records and lipsticks. Other than a rare dinner party, these simple items were the extent of IIAPCO's "graft." The industry, of course, had a hard time believing this. Some in it still don't.

Draft 3 was well received by General Ibnu, but a cable from Fowler made it necessary for IIAPCO to request a fourth draft. Permina had approved the wording in Draft 3, and I felt that if we started requesting all these new changes, Permina, too, would seek more changes.

I was aware of the general's distaste for long contracts. After a careful review of Fowler's suggested changes, we altered little other than a clause assuring IIAPCO of a favorable U.S. tax position. Ibnu was still intent on closing by August 17, Indonesia's Independence Day and a big national holiday. Japex had ten people working, trying to catch up with IIAPCO. They had finished their first draft, a contract calling for far greater expenditures than IIAPCO's. If they signed first, I felt we would lose our advantage.

On August 2, I submitted a letter to the Permina Board of Directors requesting permission to submit a final draft for signature. Two days later, I happened to meet General Ibnu at the hotel. This was the first time I had seen the General since our meeting July 4. He invited me to join a tour of South Sumatra oil fields that Permina had organized to acquaint Indonesian newsmen with the oil industry.

Except for the Sumatra press orientation trip, I stayed close to my room to be sure I was always available. I again became a prisoner in the Hotel Indonesia, the only modern hotel in Jakarta. It was only 10-20% occupied, so I knew all the guests and had the menu memorized. In spite of the exotic Indonesian fruits and exciting local foods, nearly all the hotel fare was a sorry version of Western dishes.

While I cooled my heels in the Hotel Indonesia, my stateside associates were also becoming impatient and nervous. Bob Fowler and Pat McDonough wanted to come and help. I didn't want anyone else now, as I felt it would only bring confusion. All we needed was patience.

Machmud would drop by Permina's office every day. Over lunch we would discuss the latest development and plot our next move. Ibnu gave the contract back to Dr. Sanger, informing him it was up to him. Sanger relied heavily on Kwa, with whom both Roger and I had easy communications. I began looking for office space and a house. I indicated to my family that we would move to Jakarta to open IIAPCO's office. The board approved the contract and Permina told me we needed only one more meeting to dot the "i's."

On August 15, I met with the negotiating team. Most additions and changes were double-talk run out to keep everybody happy. We made one substantial alteration. Permina asked for and I tentatively agreed to an increase of IIAPCO's expenditures by 33 $1/3$%, raising the total from $5,000,000 over five years to $7,500,000 over six years. Due to slow communications, it would take more than a week to receive approval from my IIAPCO associates.

The increased expenditures would not begin until the third year, thus retaining IIAPCO's early low-expenditure window to complete its program financing. As a trade, IIAPCO received the right to withdraw after the third year or at the end of any subsequent year thereafter. This trade-off succeeded in reducing IIAPCO's $5,000,000 five-year commitment down to a firm three-year commitment of $2,100,000. As the geology of the area was still largely unknown and Indonesia's political situation was still in a state of flux, I felt this reduction was extremely significant.

On August 18, Machmud and I arrived at Permina's office to discuss a few last changes. About 9:30 a.m., Kwa informed us that Ibnu was very upset with the delay. He was ready to sign and wanted the final draft in his hands by 1:00 p.m. Kwa asked if I was ready to sign. I was! To understand Indonesia in 1966 would be to understand why I had to sign when the opportunity arose.

Kwa and Machmud, with three flustered Permina secretaries in tow, proceeded to draw the final contract. I went into a room by myself, and with nothing to read, chewed my fingernails. We did not make it by 1:00 p.m., but we were done by 2:00 p.m. I was hustled into a car and taken to Dr. Sanger's home. He handed me a pen and I signed. He then took the papers and headed for the golf course to meet the General. Later, Andrew Alexeiev, Ibnu's golf partner, called to congratulate me. Ibnu told him he would sign the following morning.

On August 19, 1966, I sent a telegram to Larry Barker. It read:
"CONTRACT SIGNED BY ME YESTERDAY AND GENERAL THIS MORNING/COULD NOT WAIT YOUR REPLY/SIGNING BECAME URGENT/USED MY BEST JUDGEMENT/HOPE YOU APPROVE/CHEERS."

IIAPCO was in business and controlled what I felt was "half of S.E. Asia." I had arrived in Jakarta on June 28, 1966. I first met with General Ibnu and commenced negotiations on July 4. The contract was approved and signed by me on August 18 and by General Ibnu on August 19.

At times, that 7 weeks had seemed like 7 months. Today, without the decisive direction and drive of Ibnu, it takes seven months to two years to obtain a contract. IIAPCO's total out-of-pocket expenditure from inception to obtaining the signed contract was less than the cost to drill a central Montana wildcat. Those "goddam Billings cowboys" had accomplished the impossible.

The contract between Permina and IIAPCO became the worldwide model commonly used today for production sharing. It provided for a 6-year exploration period with a proviso to extend it to 10 years. IIAPCO's contract provided for a life of 30 years from the date of approval (recently extended for an additional 20 years). Twenty-five percent of the area was to be relinquished at the end of 3 years and another 25% at the end of 6 years. It provided that "IIAPCO shall designate and Permina shall approve the shape and size of each individual (relinquished) portion." Each acreage drop could be delayed by a bonus payment.

Provisions were included to allow for IIAPCO's transfer of an interest subject to Permina's approval, "which consent shall not be unreasonably withheld." The contract provided for the right to freely lift IIAPCO's share of crude. It stipulated that Permina's share of any crude sold would be booked at the "net realized" wellhead price, not at the fictitious "posted price" used by the Middle Eastern producing countries.

Permina, although responsible for the management, actually turned it back with the statement, "IIAPCO is responsible for the Work Program." Permina was to be responsible for paying Indonesian fees, charges, and taxes on imports. Permina was to make rupiah funds available to IIAPCO for local needs. The contract provided for IIAPCO's recovery of operating costs out of 40% of the oil produced. The remaining oil was to be split 65% to Permina and 35% to IIAPCO. Although IIAPCO was subject to Indonesian income taxes, "The portion of Crude Oil which Permina is entitled to take...shall be inclusive of all income taxes payable to the Republic of Indonesia."

One clause in the contract provides that IIAPCO make available to Permina, at cost, a share of Indonesia's "Domestic Requirement."

This share relates to IIAPCO's percentage of total Indonesian production. Indonesian crude was selling for $1.21 per barrel in 1964 and was only $1.56 in the summer of 1966. Rather than establish cost as a percentage of crude price, as should have been done, we arbitrarily established cost at $.20 per barrel with no provision for escalation. No one envisioned the rapid increase in crude values and costs that began in the early 1970s. This cost factor became standard in later contracts. Although as a sovereign nation, Indonesia later improved its percentage take with the stroke of a pen, it was impossible for foreign contractors to overcome this serious defect.

The contract was only fifteen legal-size sheets, exclusive of exhibits. Time has proven it a great contract, fair to both sides. It resulted in the "partnership" Ibnu desired for Permina and Indonesia. I was thankful that Harold Hutton and Tom Brook, both good independents, were responsible for acquainting Ibnu with the oil business. By the time I got to him, Ibnu already knew how independents worked, thought, and obtained their financing.

I remained in Jakarta, waiting for the Indonesian translation of IIAPCO's contract. Machmud and I continued our search for office space and housing. We made many protocol visits to Indonesian officials and at the American Embassy. I left for home with the signed contract on August 26, 1966. My associates surprised me by arranging for my wife Nancy to meet me in San Francisco. Doug Carver gave us a key to a plush Nob Hill apartment complete with champagne.

Roger Machmud proved to be a real asset to me. His background was a microcosm of Indonesian colonial history. Machmud's mother was Dutch. His father, an Indonesian doctor, was from the Sultan of

Medan's family. Roger spent most of his youth trapped in Holland with his family during World War II. He was educated in Holland and received his law degree there.

Machmud was a Shell-Indonesia employee when Permina acquired that company's assets. He joined IIAPCO when he was 32, and was paid $100 per month through our summer of negotiations. He received a $1,000 bonus when IIAPCO obtained the contract. After signing, and prior to our opening Jakarta offices, he received $250 a month. After operations began, this was increased to $750. Machmud became such an asset that when operations were turned to ARCO, Sinclair's successor, in 1968, Roger remained with the Offshore Northwest Java operation. His job description continued to expand. In 1983, he became President and Resident Manager of Arco Indonesia. Today Machmud supervises 1800 employees.

The Permina negotiating team had felt I needed legal help on my side of the table. Through the first contract draft, Machmud was on loan. When Roger was at odds with Colonel Pattiasina and was told he would be transferred to Irian Jaya, he resigned. Kwa then "pushed" his friend onto IIAPCO. I hesitated only because I did not want to come between Roger and the Colonel. Kwa assured me that this was Indonesia and that Colonel Pattiasina wouldn't be offended by IIAPCO's hiring Machmud. I doubt that IIAPCO could have obtained a contract in such a short time span without him nor could we have as easily opened an office and commenced operations. He became my eyes and ears.

Bob Fowler's legal and tax advice had been helpful in spite of the difficult communications network. Larry Barker's primary effort up to this point had been working with Fowler and seeing that my handwritten letters were typed and sent to our partners. His work load and input escalated through the ensuing months and years.

IIAPCO knew that with its associates it could financially handle the start-up period and the most preliminary exploration effort. IIAPCO, shared equally by McDonough, Barker, and Todd, owned 65% of this new contract. Carver-Dodge owned 25% and Warrior 10%. Carver-Dodge and Warrior, although small independents, had superior financial resources.

Larry Barker, generally cash short, represented strong assets. Pat McDonough, the original credit card millionaire, controlled a small well-completion and workover company. (Throughout his life, Pat originated a number of great ideas, but he lacked sufficient resources and patience to follow one idea to fruition before embarking on the next. His IIAPCO ownership decreased on several occasions as he sought financing for a new deal. In 1966, McDonough assigned half his IIAPCO stock interest to Chesley Pruitt, an Eldorado oil man with deep pockets.)

PRESIDIUM APPROVAL

On August 30, *Platts Oilgram* announced IIAPCO's contract signing and it ran follow-up articles on September 1 and 9. The oil industry, including all the majors, was now aware that IIAPCO had signed a Production Sharing Contract with Indonesia. A whole new game was to unfold. We needed Presidium approval and we expected it within a few days of contract signing. We were to spend a frustrating five months waiting for it. A number of major oil companies attempted to have our contract and three subsequent contracts voided by pressuring the U.S. Embassy and the new Minister of Mining not to support them.

When I started negotiations, General Ibnu was Minister of Mining, Director General for Oil and Gas, and President Director of Permina. When it was deemed Ibnu should not hold all three positions, he elected to give up his position as Minister and remain Director General and President of Permina.

On July 25, Bratanata was named Minister. On August 30, Bratanata informed Ibnu that Marshall Green, Ambassador Jones's successor, said there was not any "big money" behind IIAPCO and implied that Permina should have signed a contract with one of the major oil companies. Ibnu replied that he had signed a contract with one of the independents on purpose—they were more progressive and would keep the majors on their toes. He told Bratanata that if the majors weren't willing to change, that the independents of today would be the majors of tomorrow.

This signaled the start of an all-out struggle for control of oil policy. Bratanata, supported by the majors and the U.S. Embassy, was trying to control Ibnu. Ibnu's decisive, resolute, single-minded approach brought opposition from Suharto's technocrats. Ibnu neither sought nor desired consensus.

It is natural that such a man would find enemies. The unions had long tried to see Ibnu removed. The World Bank and the IMF were opposed to Ibnu committing to large loans on behalf of Permina without going through central controlling authorities. Ibnu was opposed because of his strength and his refusal to accept governmental controls.

Machmud told us that although we had anticipated quick Presidium approval, it might now take a few weeks. Presidium approval was essential because IIAPCO's contract was not official without it. Sutan Assin would not allow the opening of Indonesia's geologic files to us until after this approval was gained.

Permina's view was that Presidium approval was not necessary, but would come in time. They felt IIAPCO should proceed. We had to gamble on Presidium approval. If I was nervous, my associates, unacquainted with Indonesian and Asian ways, were even more so. It was difficult not to jump to premature conclusions, and it was justifiably hard for my partners to trust Machmud's judgment and mine.

Bratanata's weakness was his inexperience. Ibnu's weakness was his assumed "record of corruption," his position against unions, his resolve, and the fact he had ties to Sukarno's regime. Rumors indicated Ibnu was losing his power. We were to hear this rumor many, many times before he was finally deposed in 1976.

In Billings and Denver, we began to prepare for a geophysical program. After Fowler confronted the U.S. Embassy on its big-company attitude, he received a letter from McCusker denying that Ambassador Green or any other embassy official had ever commented on IIAPCO's financial capabilities to Bratanata. There was a possibility the embassy was being used in this struggle.

My personal relationship with embassy personnel and the help I had received from them had been excellent, so this intrigue, although fascinating, was disconcerting and disruptive. To put my mind at ease, Machmud sent a cable: "INTRIGUE NATIONAL SPORT HERE. TAKE SEDATIVE."

Kwa and Dr. Sanger were confident that General Ibnu would obtain Presidium approval and advised us to relax and show utmost confidence in him. But his frequent business trips out of the country left a vacuum. It became a time for attacks by his detractors.

The basic opposition of the large oil companies to the IIAPCO contract related to management clauses. On the face of it, Permina had that responsibility. (In the 25 years since IIAPCO first accepted Permina's management participation, no company has found any serious problem in the partnership arrangement created by it. Our faith in Permina and Indonesia was justified.)

Bratanata was still not convinced that Ibnu had no interest in IIAPCO. Tenneco, among others, suggested to Bratanata that the IIAPCO group did not have the capital to implement its contract. Bratanata made it clear he would deal only with foreign parties introduced by their respective embassies. We badly needed U.S. State Department support for our cause.

McDonough, Barker, and I went to Washington to meet with Ambassador Green and McCusker in an endeavor to seek embassy support. We first met Senate Majority Leader Mike Mansfield, a former professor of Pat's, who arranged the meeting with Ambassador Green. The ambassador indicated he would try to counteract the unfortunate remarks attributed to him by Bratanata.

I had to return to Indonesia to meet with Permina officials and to personally meet Bratanata. We had no intention of betting on two horses. We merely hoped to lighten Bratanata's suspicion and stand up against our stone-throwing competition. It was Bratanata who was not passing the contract on to the Presidium for approval. It was up to IIAPCO to prove we were not "Independent Ibnu American Oil Company." (I received word from Machmud that he was broke because transfers IIAPCO made took forever to reach his account.)

In late September, while Ibnu was out of the country, Bratanata gave General Suharto a report outlining all Ibnu's alleged sins. Ibnu declared war. He told his staff they could choose Bratanata or him. "As far as I am concerned, there is no longer a Minister of Mining," he said.

Ibnu signed an IIAPCO-type contract with Japex on October 6, 1966. On October 7, Ibnu was called to General Suharto's office to confront Bratanata regarding the accusations made against him. Ibnu reportedly refused to shake hands with Bratanata. A number of generals appeared to be on Ibnu's side, including Suharto.

IIAPCO was walking a tightrope. We were in Ibnu's corner, yet we had to put Bratanata's suspicions to rest. Bratanata had requested credit references on IIAPCO and its associates through Bank Negara Indonesia. Politics was interfering with the preparation of our work program and budget. Work at my map table became even more remote.

The Japanese Embassy gives strong support to Japanese business, in sharp contrast to the U.S. Embassy's hands-off policy. Because of this, the signing of the Japex contract gave IIAPCO encouragement that eventually both contracts would be ratified by the Presidium. Paul Cleveland, Commercial Attache and my closest embassy friend, was thoroughly enjoying this slugging match. He felt the winds were shifting in Ibnu's favor. His information was that Ibnu was now dealing directly with Suharto. He also reported one major oil company's complaint that it looked as though Ibnu was closing Indonesia's offshore to them.

On October 15, Pan-American made an application for an area that overlapped 25% of IIAPCO's contract area. Pan-Am suggested to Kwa they share seismic and other costs with IIAPCO on the overlap area. Later Pan Am claimed the overlap was an oversight.

Union broke off negotiations when Permina refused to allow them to make some desired changes in the IIAPCO type contract, a contract modeled after an early Union draft. Refining Associates (Refrican) immediately picked up where Union had left off. Refrican signed a contract on October 20. Obviously, Ibnu intended to quickly sign a number of IIAPCO-type contracts to force the issue with Bratanata.

On October 20, Bratanata sent a communique to the Presidium on the subject "Cooperation with IIAPCO." He stated that "While Ambassador Marshall Green once said that IIAPCO was not well known, on October 16, His Excellency had amended the remark, adding that IIAPCO included five companies technically and financially sound." Our efforts were beginning to pay off. In the same letter, however, Bratanata referred to "the IIAPCO draft contract" and requested we be called for final discussions. He proceeded to outline several suggested changes, including discontinuing the name, "Production Sharing Contract."

I left for Jakarta October 25, 1966, to present IIAPCO's 1967 Work Program and Budget. Most importantly, IIAPCO felt I should be there to get a first-hand feel for what was going on. Caltex was becoming frustrated waiting for approval of their five-year Central Sumatra expansion program. They were also faced with making expenditures without formal approval. Ibnu had asked Caltex not to even see Bratanata.

Julius Tahia, president of Caltex Indonesia, guided their policy. As an Indonesian, he had a better feel for the Bratanata-Ibnu struggle than most foreigners. I was advised that IIAPCO should proceed cautiously, as if we actually had Presidium approval. This was not easy to sell to my associates. I was told the Jakarta oil fraternity felt IIAPCO had a contract because we had begun negotiations before Bratanata appeared on the scene.

At home, IIAPCO was also busy choosing up sides. Barker was spending most of his time on IIAPCO budget planning. We were getting great support and help from Carver-Dodge. Doug Carver was the biggest of IIAPCO's "butter-and-eggs" partners, and a great person. I felt he would be a big help in establishing IIAPCO's financial credibility. Will and Joe Obering had become part of the team.

Everyone except Pat McDonough agreed on almost every item. Pat began to vote indiscriminately against everything. He felt Larry and I were siding with Carver, with whom he did not get along. Pat had contributed neither financial nor moral support. After I'd been playing politics in Indonesia and the U.S. for three years, Pat told me, "stick to your drafting board; I'll take care of the politics." Pat was frustrated not to be calling the shots, even as he declined to be an active participant.

In 1947, I had had a summer job roughnecking in northern Montana. Pat had been the other floor hand, earning summer money for school. When I moved to Montana in 1951, I found Pat in Billings. For many years he was a next-door neighbor. Our children played together. Pat had his office just down the hall from mine. This close friendship of 20 years was beginning to fade with his opposition to all of

IIAPCO's efforts.

In Jakarta, I found the Hotel Indonesia full of foreign businessmen. In two months, local prices had increased 50%, some even more. Coffee, which had been a dime, was now 40 cents a cup, and a 25-cent haircut was now a dollar.

IIAPCO and its ratification problem had become an open topic. I felt the U.S. Embassy was our worst enemy. They did not understand Ibnu's cold, efficient way of business, and they knew they would never sway him, while they felt they understood and could influence Bratanata. Sixteen oil companies had been in Jakarta in my absence. Most, introduced by Ambassador Green, chose to deal with Bratanata. Major oil company brainwashing of embassy personnel influenced U.S. Embassy position even though the staff truly believed they were remaining impartial.

It was difficult, if not impossible, to convince people that Ibnu had no interest in IIAPCO. Shell checked IIAPCO out and reported to the embassy, "They're just five little companies," and instilled the notion IIAPCO would quickly sell out.

My job became one of educating the U.S. Embassy and Bratanata on the function of small independents, IIAPCO in particular. I invariably carried my geologic maps to meetings. I would unroll them on the floor and endeavor to explain the geology of Indonesia and IIAPCO's area. I tried to explain how independents financed projects by taking a number of partners, in contrast to floating a new issue of stock, as a large company might do.

At Permina, IIAPCO was a hero for working out the contract that was bringing in oil companies from around the world. On October 31, Suharto gave Ibnu direct access to the Presidium. Ibnu was back on top! I had the pleasure of breaking this news to the embassy. Few outside of the Permina inner circle had given Ibnu and IIAPCO much of a chance.

What fun I was having! Machmud deserves enormous credit for not allowing IIAPCO to get cold feet. Ibnu informed me that Suharto's advisers thought the IIAPCO contract was a great contract and that only Bratanata had been against it. He told me we could now get started. However, we still didn't have the signed Presidium document.

I met Bratanata, finding him pleasant and easy to talk to. There was no talk of cancellation or renegotiation of IIAPCO's contract. He only wanted to ask about IIAPCO and those behind it.

Doug Carver arrived on November 5, in time to accompany me on a trip with General Ibnu to a North Sumatra spud-in ceremony. Ibnu took every opportunity to sit with us to demonstrate that IIAPCO had his full support. This trip was so successful in helping me to become acquainted with Ibnu, Permina, and other Indonesians, and to see more of Indonesia, that I invited myself on many more dedication trips.

On November 8, 1966, the *Singapore Straits Times* carried an article entitled, "The Man of the Hour." It said that apparently General Ibnu and three "lightweight" independent oil companies had won the fight against Bratanata and 16 companies seeking contracts in Indonesia. Oil company and diplomatic sources were quoted as saying this would be to the eventual detriment of economic growth of Indonesia. Obviously our struggle was not yet over.

On November 10th, we had a long meeting with Ibnu. He discussed the current political game that continued to be played by the majors through Ambassador Green. He told us he would see that the IIAPCO contract was approved by the Presidium. He said he didn't care where IIAPCO got its money for its program. He said, "Just do your job and don't embarrass me." He suggested that on our return to the U.S., we go to Washington and ask our State Department to get off his back, to quit interfering with internal Indonesian politics, and to quit choosing sides.

During the meeting with Ibnu, I mentioned IIAPCO's house-hunting problem and my dismay at the thought of staying in a hotel with my wife and five children for any prolonged period. Unknown to us, Ibnu asked Hadji Thahir, an outside business associate, to find IIAPCO a house. Permina purchased a house, which Thahir indicated was an Ibnu decision not to be questioned by IIAPCO. It appeared to be a case of those under Ibnu exploiting the situation. Ibnu later indicated IIAPCO was not obliged to take it, that choice of a home and office was entirely IIAPCO's. Permina converted this home, purchased with IIAPCO in mind, into the first Indonesian Petroleum Club building.

I found Ibnu to be very fair and honest, but that did not prevent some of those under him from trying to take advantage of business situations. When similar situations arose during IIAPCO's start-up period, I went directly to Ibnu. In every case, he stopped those within Permina from misinterpreting the contract for personal gain.

Machmud and I had looked at a large house suitable for a home as well as an office. However, Mussry, the owner, requested that payment be deposited in Hong Kong in U.S. dollars with no receipt given to IIAPCO. We eventually bought this house under these difficult conditions after personal clearance and approval of General Ibnu.

For nearly a month I had been so busy in Jakarta that I had not written home. On boarding the plane for the United States, I was handed a cable reading: "IF STILL ALIVE PLEASE INDICATE. IF NOT NEVER MIND. NANCY"

Upon our return to the U.S., Carver contacted Senator Arthur Younger, who arranged a meeting with Undersecretary of State William Bundy in Washington. We stirred up a bit of a hornet's nest for the Jakarta Embassy. The embassy continued to claim it was a misunderstanding, reporting to Bundy that they were all for IIAPCO, disavowing Bratanata's contention that IIAPCO was "small fry."

Barker, Machmud, and I constituted IIAPCO's entire staff and were now all running double time. We were endeavoring to set up an exploration program, work on a Permina Operating Agreement, look for a Jakarta home and office, seek insurance coverage, do market research, keep our partners and Permina informed and happy, plan additional financing, all the while preparing to move me with my family to Jakarta. And we were still nervously awaiting Presidium approval.

Kyushu signed an IIAPCO-type contract on November 22. It committed them to a $25,000,000 expenditure over eight years and $15,000,000 in production bonuses. This compared to IIAPCO's commitment of $7,500,000 over six years and no production bonuses. Now, in addition to IIAPCO's contract, offshore contracts had been signed with two Japanese companies and one Canadian firm.

IIAPCO's contract was presented to the Presidium, but was held in abeyance to be sure it did not conflict with the new Foreign Investment Law currently being debated in Parliament.

Indonesia passed the new liberal law in late December. As it was now Ramadan, the major Moslem holiday, we could not expect ratification before mid-January.

On December 20, 1966, Bratanata sent a letter suspending the Japex, Refining Associates, and Kyushu contracts. This was followed on January 7, 1967, by Ibnu's letter to Suharto stating that as Director General of Oil and Gas he had approved offshore contracts with IIAPCO, Japex, Refining Associates, and Kyushu and requested a prompt approval by the Presidium.

Soon after Ibnu's letter was sent, Japan applied strong diplomatic pressure on Indonesia to ratify those contracts signed by General Ibnu. This effort by "Japan Incorporated," supported by the Canadian Embassy, resulted in Presidium approval of all four contracts, including IIAPCO's, on January 19, 1967. The U.S. Embassy chose not to become involved.

This approval was followed by a directive putting the Director General for Oil and Gas under the direct supervision of Suharto. Those "Billings cowboys" had overcome some tough opposition. It was now apparent that after IIAPCO first put pressure on the U.S. Embassy, our contract was not among those "suspended" by Bratanata.

The *Oil Daily*, February 7, 1967, carried an article reporting Suharto's approval of four firms, including IIAPCO. It said that the announcement crushed the hopes of at least 15 other foreign companies, including Standard Oil (New Jersey), Mobil Oil Co., Phillips Petroleum Co., and Gulf Oil Corporation, which were bidding for the same sites.

I arrived in Jakarta to this good news in late January, 1967. Ibnu approved our purchase of the Mussry house. My family was scheduled to arrive in mid-February. The hotel was full. Student anti-Sukarno demonstrations were in full bloom. I was gearing up to start spending. My only worry continued to be money. We needed money now!

I went to Singapore in early February to open bank accounts and to purchase everything we would need to open our house and office. I had arranged for the use of Permina's new Fokker Friendship to fly everything to Jakarta along with my family. I hoped to be in our new house and office by the first of March, ready to do business. I was concerned with spending and committing to large sums knowing how short IIAPCO was of funds. Carver told me not to worry, he and Barker would figure it out.

My family arrived in Singapore on February 14, 1967, happy, but a bit bewildered by it all. My wife Nancy was delighted to relinquish her role as cruise director to me. Nancy and our five children, ranging from 8 to 15 years of age, had never before traveled beyond the U.S. and Canada.

On February 17, all our Singapore purchases except for a car and piano were loaded on the plane, and we took off for our new home. Machmud met us at the Jakarta airport. After clearing customs, he found an army truck on the street, which he "hired." Within three hours of landing we had cleared customs and had our furniture and equipment unloaded at our home!

A geological consultant from Manila was engaged to assist in the programming, selection and supervision of IIAPCO's early geophysical program. We had a difficult time finding a secretary who could take English dictation and type it correctly at a speed exceeding eleven words a minute.

FINANCING

In November 1966, Doug Carver discussed IIAPCO's contract and its ratification problems with Ralph K. Davies, Chairman of Natomas Company, a San Francisco shipping and real estate company listed on the New York Stock Exchange. Prior to purchasing control of Natomas, Davies had formed American Independent Oil Co.

(Aminoil), one of the first U.S. independents to become involved in foreign exploration. Aminoil's start-up problems—finance, as well as major oil company opposition—were like those currently faced by IIAPCO.

Davies had been a marketing vice president at Standard of California (SoCal) until he became Interior Secretary Harold Ickes's right hand in the Petroleum Administration for War (PAW) during World War II. His return to SoCal after the war was resented by his replacements and left Davies with a chip on his shoulder for major oil companies, and SoCal in particular. Now Davies felt IIAPCO needed a financially strong and prestigious industry partner and he saw this as a chance to help another underdog.

As a result of Carver's meeting with Davies, and with the approval of the IIAPCO Group, participation discussions were opened in late December with Signal and Aminoil. Natomas would join if these companies found our proposal attractive. This group was given until February 15 to study our proposal. At that time, Pat McDonough's dislike of Doug Carver caused him to be irritated by Doug's active involvement in the IIAPCO Group's financing effort.

Signal-Aminoil declined to join our venture. Davies, however, still interested, arranged a New York meeting with Sinclair, a company with which Natomas had business dealings. On Monday, February 27, 1967, Barker, Carver, and Ben Tyran, executive vice president of Natomas, met with Fred Bush, president of Sinclair, and his top technical and legal counsel. A letter of intent was signed on Friday, giving Sinclair an exclusive option for 60 days (later extended to 90) to study, evaluate, and write agreements. Meanwhile, Barker sent the IIAPCO co-owners a meeting notice saying, "WE NEED MORE MONEY IN THE IIAPCO BANK ACCOUNT." This project was now for real and was costing real money.

The trade offered Sinclair would give them a 51% controlling vote. Sinclair would own 46% of the contract. Natomas would acquire a 5% ownership and give its vote to Sinclair.

The expenditure split proposed was:

Expenditure	Sinclair/Natomas	IIAPCO
1st $500,000	90%	10%
2nd $1,000,000	85%	15%
3rd $1,500,000	80%	20%
4th $2,000,000	75%	25%
Thereafter	51%	49%

Sinclair also committed to a $9,000,000 IIAPCO development loan.

I was asked to return to New York in March for conferences with Sinclair. They needed me to discuss geology and to give my impression of the political climate in Indonesia, to discuss fluctuating rupiah exchange rates, and to try to convince Sinclair of the validity of IIAPCO's contract. Sinclair wanted assurances that Permina would allow us to take partners, and they wanted to know more about how the Work Program functioned.

Back in Jakarta, I had left my family in a completely unsettled situation. They had been in this strange foreign land for only three weeks.

Two doors away, former First Deputy Premier Subandrio's trial was taking place. The military set up a machine gun in our driveway. On one occasion, armed soldiers stopped IIAPCO's car with Nancy and the children in it.

My family survived this indoctrination period in great shape, but much more lay ahead. When I was away, besides managing five kids and ten servants, Nancy was responsible for IIAPCO's Jakarta office. Machmud kept her informed and she signed the checks. A large company would have had at least ten expatriates for a start-up. Nancy and the kids were learning a few words of Indonesian and were now able to communicate to a limited degree with the driver and household help. They were unpaid IIAPCO employees.

We still couldn't get Mussry to move out of the first floor, so it was impossible to start the necessary home-office remodeling. Mussry refused to budge until he was paid. We didn't want to make the final payment without a receipt. Mussry wouldn't give a receipt as these funds were being deposited and hidden in a Hong Kong account. Still, IIAPCO needed to account to Permina for expenditures. Impasse! We needed to give Permina $50,000 before they furnished us with the necessary rupiah for local purchases. We had fires going everywhere with only Machmud and Nancy in Jakarta to put them out. Our car arrived, but was tied up in customs.

On Monday, March 17, 1967, Sinclair furnished IIAPCO a first draft of the agreement and its accounting procedure. Sinclair continued to show enthusiasm for our venture. The one big holdup in final approval remained Sinclair's concern over whether Indonesian law required IIAPCO to have parliamentary approval rather than just Presidium approval. E.L. Steiniger, Chairman and CEO of Sinclair, was not as concerned with legality as was Sinclair counsel. Steiniger viewed this as a major opportunity to become more self sufficient in crude. Sinclair had to purchase 285,000 barrels of crude a day to meet its refining and marketing needs. By the end of the week, we were in agreement on most issues.

While I was still in New York, the political crisis in Jakarta eased. Nancy wrote, "The guns are all gone now and we haven't seen a tank in days." Indonesia jacked up the price of gasoline 800%, to 16 cents per gallon. Sukarno portraits were being removed. After early March, General Suharto alone held power, thus ending dual leadership. President Sukarno had been eased out in this Asian way. Civil war had been avoided.

The press reported that General Ibnu had been arrested on charges of corruption. It was alleged he owned 11 private cars and drew a salary of 1,500,000 Rp per month—$4,500 at the official rate, $1,500 at the black market rate. He was charged with misappropriation of funds to indoctrinate his workers against the oil companies. Perhaps it was thought he had a grudge against the majors.

Once a young Indonesian student came to our home seeking financial support for his "anti-corruption" cause. He had no idea that I was associated with Permina, his principal "corruption" target. When I asked how he knew General Ibnu was corrupt, his reply was, "God strike me dead if I'm wrong."

By the end of March, Bratanata was again raising his head. The *Oil Daily* reported a Bratanata statement saying Parliament still had to act on four offshore contracts signed by his subordinate Major General Ibnu Sutowo. *Platts Oilgram* reported that Bratanata had solicited eleven oil company proposals for offshore Kalimantan. This was all hitting the press as Sinclair was trying to convince themselves of the legality of IIAPCO's contract.

In early April 1967, the Seventh World Petroleum Congress was held in Mexico City. Both Mr. Steiniger, CEO of Sinclair, and Pat McDonough attended. Pat introduced himself, blew out his chest, and told Steiniger the IIAPCO deal was too big for Sinclair. Pat implied that IIAPCO was visiting with Pan American at the same time Sinclair held an exclusive letter of intent.

This unfortunate and inappropriate bragging by Pat, who had not talked with anyone, combined with Bratanata's last stand to cause Sinclair to consider withdrawing. Carver hopped a plane to London, where Fred Bush was on business. Carver was able to clear up the misunderstanding and get the negotiations back on track. Texaco, on learning of our discussions, tried to convince Sinclair not to become a partner with IIAPCO.

Bob Elston, president of Sinclair International, was not involved in any of the March meetings, and was very wary and cautious. Negotiations had been stalled for three weeks. The May 6 closing date was extended for 30 days so that Sinclair could send a team of technical and legal people to Jakarta.

About three times a week, I'd call on Permina just to keep IIAPCO's face in front of them. Machmud went every day. I met with Bratanata. He still thought he could change the contract. I informed him we were prepared to implement the contract as written.

By late April, anti-Chinese demonstrations were common. Although only 15% of the Indonesian population was Chinese, they reportedly controlled 85% of business. The Chinese did not mix with Indonesians, even though their families might have lived in Indonesia for generations. At the children's school, no Chinese now attended. Unfortunately, this included half of the teachers as well as half the student body. There was talk of deporting the Chinese.

Larry was trying to convince both Sinclair and IIAPCO's partners that U.S. legal concepts were based on English law, while Indonesian law was interpreted from an oriental–Dutch colonial viewpoint. Indonesia was an Asian nation that had been independent for only 18 years. I had worked very hard to understand and accept a culture and mentality completely different from the one in which I had been raised. Larry and I had worked together for so many years he had no problem accepting my analysis, but it was not so easy for us to convince others. We had to be sensitive to Indonesian perspectives. Permina was beginning to be upset at our constant nagging about legal title.

In mid-May, 1967, Carver and Barker, along with Bernie Moore, Ray Halsey, and David Waer of Sinclair, arrived in Jakarta. Sinclair's team dug into everything legal and technical. Barker, Carver, Machmud, and I endeavored to answer their questions. We met with Permina. Sinclair had satisfied themselves by month's end, when Steiniger and Elston arrived. Elston remained nervous and hesitant. On June 1st, at the end of two days of meetings and discussions, Mr. Steiniger asked Bernie Moore what he thought. Bernie answered, "drill it!" and Steiniger turned to IIAPCO and said "It's a deal. We are indeed partners. We should proceed."

That afternoon, Steiniger met with Ambassador Green and informed him of Sinclair's commitment. That evening, we had the first big dinner party at our new home. Our handsome handmade teak table and chairs were used for the first time. General Ibnu and his wife Sally were honored guests along with Mr. and Mrs. Steiniger.

The following day, Elston, Barker, and I met with each of Suharto's Presidium ministers to pay our respects and to introduce Sinclair. We first called on Minister Sanusi, who told us *not* to start physical operations until we had contract approval from Parliament. This shocked the already-wary Elston, who then suggested to Steiniger that Sinclair renege on its commitment. Steiniger told him a deal was a deal; for better or worse, Sinclair and IIAPCO were partners. Elston was finally behind us. It had been a long road. On June 3, 1967, Sinclair, Natomas, and IIAPCO issued a joint press release.

FAMILY LIFE

On February 17, 1967, the night my family arrived in Jakarta, I took them to meet General Ibnu and his family. Then we went to the Glodok, Jakarta's Chinatown, for a Chinese dinner. Rats in the alley and cats under the table distracted my family from the superb food.

One week after their arrival, we all moved into our new home—although Mussry still occupied the lower level. A local cabinetmaker was building our furniture from sketches I made. His "factory" had no electricity. Even the lathes were foot driven. We started housekeeping by hiring a cook, driver, houseboy, and wash lady. All laundry was done by hand on rocks. We had a kerosene cook stove. Nancy was naturally concerned for the children, but we were all having a ball in a new and unbelievable environment.

Major Judo Sumbono, wearing a holstered pistol, came to welcome us to our new house. Relaxing with this Permina friend, I put my feet upon the table. Later, I learned it was an insult to expose the soles of one's feet to a Muslim. He was most upset with me. Nancy sent word through Machmud that *she* was upset that Judo wore his pistol into our home. We were all learning.

As none of our original household help spoke English, and we had not yet mastered Bahasa Indonesian, we encountered a few problems. There were only 90 American families living in Jakarta, most of them attached to the U.S. Embassy. There were no theaters. Entertainment was essentially home entertainment. Because the American community was so small, we were allowed to join the American Club, an embassy-sponsored establishment with a great swimming pool. We enrolled the children in an English-speaking school run on the British system by an order of Roman Catholic nuns, most of whom were Chinese.

At the end of February 1967, we drove to the mountains, where we had a chance family meeting with President Sukarno at a tea house. After visiting with the children, he insisted on meeting their mother. The following week, Sukarno was removed completely from office.

We rented a dilapidated boat and went out to the Thousand Islands to snorkel and get sunburned. One time it was a Caltex boat, its crew moonlighting for a few extra rupiah. We went to the beach. We took a 480-kilometer (300-mile) death drive to Bali, where our 12-year-old son had his first view of bare breasts—at least that we knew of.

We climbed into volcanoes. We rode the Bimah (Blue) train to Jogjakarta to visit Borobudur, the 9th-century Hindu ruins. We went to Samudra Beach on the Indian Ocean and to the botanical gardens at Bogor. We watched batik and candlemaking and we bought works of art. Once the house was operational, we allowed the tukangs (street vendors) to come each evening at cocktail hour. We bargained with them for antique china, cannons, old bronze, and prehistoric adze heads. We seldom spent more than five dollars in one of these bargaining sessions; most pieces did not exceed two dollars. The man from the gold district became our friend. Although poorly lit and layered with dust, the Jakarta Museum was great.

The tropical fruits of Indonesia were wonderful. Orchids grew even without a green thumb. On Sunday mornings, our oldest daughter often went to the flower market to bargain for a big bouquet of flowers.

Once a week we invited a Permina family to dinner. We always served them home-made ice cream (made with powdered milk and canned cream), and they loved it. We made friends. Our oldest son's best friend owned an orangutan. The change was dramatic, educational, and great fun for all of us. No one missed T.V.

We were given two dogs, who proved very useful. The dachshund was adept at killing large cockroaches. The boxer was able to jump rope with our youngest daughter, as well as open all the doors in the house.

It took fourteen months to get the swimming pool operational. It was necessary to boil our drinking and cooking water. The cook felt that was not necessary for ice cubes, but it really is! It seemed for a time that our refrigerator wouldn't even make ice cubes. Then we realized the cook and servants were making a little extra money selling them. There was no pasteurized milk. We rinsed the lettuce with Lysol.

Once a week, the "Bandung" man brought vegetables from a farm 160 kilometers (100 miles) away. The veal man came by once a week on a bicycle. He would slice a chunk of meat off for us on the spot. There was an egg lady and a bread man. I hand-carried goodies such as jello, peanut butter, and powdered milk in from Singapore. The servants loved our cocoa— and anything else we imported. One day our nine-year-old came home from school all sweaty. When I asked if he was hot, he replied, "Daddy, I'm always hot!" That's the best description of Jakarta's climate. Every afternoon a cloudburst would flush and clean the drainage ditches and canals, which were Jakarta's open sewer system.

The electricity—when it was working—fluctuated wildly. Our fuse box was a creative piece of artwork. We thought we didn't have enough hot water in the house until we found the hot water was hooked up to the toilet system. Our water well carried an excellent rainbow of oil—at least a good sign for IIAPCO's effort. Our cook, cleaning pots and pans with a pumice stone, took all the copper off the pan bottoms. One night, while I was away, a rat made its way onto Nancy's pillow. Her scream brought the servants on the run and frightened the rat. We covered the open drains with screen to slow the entry of the cockroaches and rats.

One weekend, I was a guest at a geologist's graduation ceremony at the Bandung Institute. While I was gone, it rained so hard the force of the storm broke holes in the IIAPCO ceiling. The house had only bare electrical wiring—no insulation. A neighbor assured Nancy that no one would get electrocuted, as Jakarta's power wasn't strong enough. After two months, Indonesian customs still hadn't released our car. It had a 24-hour armed guard to prevent items from being stolen off it.

By May 1967, we finally had a telephone. It had taken two months to get a telephone number and an additional two months to get a connection. Now it worked for local calls, at least 50% of the time.

After having a plumber out eleven times in four and a half months, our tub and shower finally worked. So much for the "western-style" bathroom that so influenced our decision to purchase that house. It was necessary for me to travel to Singapore once every two weeks for supplies, meetings, banking, and to call the U.S. Once Nancy came to pick me up at the airport and she met six planes before mine landed. I had sent a cable advising her of my arrival time, but the cable didn't arrive until four days later.

After Sinclair became a partner in June 1967, Carver and Barker insisted I take some time off and get away from the pressure. When everyone had gone back to the U.S. and the kids were out of school, we took ten days off and drove to Bali. Of course, once I was not available, everyone in the U.S. wanted to know what the Hell I was doing taking time off when so much had to be done!

Once the house and office were operational, we settled into a routine. Breakfast at 6:30 a.m. Kids off to school at 7:00. At 7:30 I would leave home in coat and tie to go to the office upstairs. Machmud called on Permina every day; I went down there two or three times a week. Driving to appointments and meetings was hot and time consuming, accomplished with horns and courage by an intrepid driver.

Our household staff reached a maximum of ten. It included two drivers, a guard, a cleaning person for each floor, two houseboys, a cook, a gardener and a wash lady. The total cost of the household staff, including wages, rice, kerosene and tips, only came to U.S. $90.50 per month. It was, however, like having ten more children. It seemed someone from our household staff was always sick, had a stomach ache, was burying a father, or had a sore toe. The entire staff loved the taste of Pepto-Bismol. The cook invariably cremated the fried eggs about 30 minutes before she served them. On Sundays, if the traffic thinned enough, I would venture short distances as driver of my own car. We had cultivated an orchid wall off our veranda.

IIAPCO's second vehicle arrived in August. Someone stole the trunk key and customs wouldn't release the car until they could see inside. With so many Sinclair and United Geophysical people coming and going, we were in desperate need of additional wheels. I also had Permina meetings and the kids had to get to school. The Chinese ethnic problem did not improve. The children's school closed in late August. We packed our two oldest daughters off to school in Denver. Our three youngest children were enrolled in the International School.

On most days in September, the temperature in this humid climate reached 100° Fahrenheit. We had an air conditioner in the office, but there was not enough electricity during the daytime to run the household air conditioners. Besides lacking fresh milk, jello, candies, and other foods and goodies we were so accustomed to, we lacked books, magazines, and newspapers. Our world grew more comprehensible when international news magazines became available.

In spite of a few shortcomings, we were having fun as a family, and business was exciting. When we needed to find a site for another Raydist station, I hired a 20-meter (65-foot) navy boat and piled Nancy, our kids, and all their friends aboard and headed for Ringit Island, 70 miles out at the north end of the Thousand Island chain. We slept in sarongs and cooked on the beach of this tiny, uninhabited island. IIAPCO's operation during my Jakarta tenure was a family affair. This was even important in our business relations with Indonesians, and Permina in particular.

We hired our tenth cook and fired the *babu* (maid) because she was having an affair with our married driver. Saturday was the best time to write my weekly letters, because no one but Nancy and I worked Saturdays. By November our swimming pool had become a joke. Ten people were working on it.

Jakarta decided to paint traffic lanes and sidewalk crossings on the streets. We counted 45 men in a $1/2$-kilometer stretch, painting stripes. Drivers thought they should aim down a stripe rather than drive between them. We found a dishonest "official" American friend to buy some goodies for us from the Embassy commissary. We had forgotten what hot dogs looked like. Our water well went dry and we had to haul water by the bucket. The servants had quit fighting!

Someone stole Nancy's typewriter. We were sure it was an inside job. The servants brought in a skinny little old man to perform some black magic and cast a spell on the guilty party. He got a bowl of water, put a magic pebble in it, stirred it with a *kris* (power sword), and chanted. Everyone, including Machmud, Nancy, and me, drank from the bowl and signed a paper stating that if the guilty one died, the seer's name would not be released. No one dropped dead or confessed and the typewriter never showed up. The one we suspected had only a headache.

Our new generators were operative. The office air conditioner had broken down about once a week due to the fluctuating Jakarta electrical current. By mid-April the swimming pool was working. Just about the time that the air conditioners and pool were operative, it was time to begin packing for our return to the U.S.

In mid-May the packers arrived. The Todd family's Jakarta stay was fast coming to an end. We had arrived with one suitcase each. We were taking home a boatload of handmade teak furniture, new and antique. We had collected 22 old cannons, a few hundred pieces of antique china and brass, a bucket full of prehistoric adze heads, stone and wood carvings, paintings, and more. We had been busy collecting some great things, especially wonderful memories.

Even after our return to the U.S., my family remained involved with Indonesia and our many Indonesian friends. Nancy frequently accompanied me on trips to Indonesia. We entertained visiting Indonesians and their families. A young Indonesian coed lived with us for seven years while she went to college, allowing us to continue to enjoy *nasi goreng* and other Indonesian dishes.

Our two oldest daughters returned to Indonesia to study the art of batik-making as guests of Ibnu and Pertamina. They paid for their trip by selling Natomas shares—at a good profit—in which they had personally invested.

Our older son spent a college summer doing field geology for Pertamina in Java and Kalimantan. Our younger son took several months between high school and college to work in Trend's Irian Jaya operation. Both sons are now geologists working the international arena, one for a major oil company and one as an independent.

Indonesia was a positive experience and influence on our entire family. It affected the way we all think and act.

THE SEISMIC PROGRAM

We needed to survey onshore Raydist station positions for seismic line control. Surveyors were scarce in Indonesia. After much effort, Machmud engaged a Mr. Neumann, a German, who claimed to have charted General Rommel across the African desert during World War II. Neumann left Jakarta loaded down with his transit and other survey equipment. He finished in late June, 1967, having been in the field for two months.

To verify the survey, United Geophysical flew a lane count—only to find a severe survey error. Machmud went to the villages in the vicinity, where the locals informed him Neumann's survey had been conducted by using compass line of sight from treetop to treetop. His "chain" had been a knotted rope. He had never unpacked his transit! We were told that he spent most of his time romancing the village chief's daughter. To correct our survey, we had to fly in a United team that had been working in Spain. This was the only serious disaster that can be attributed to IIAPCO's short Jakarta staff.

IIAPCO hired a superb draftsman by the name of Bachtier. He joined the Arco payroll with Machmud when operations were turned over to Sinclair a year later. I had hoped to spend more time accumulating and posting geologic data. Soon we would need to start tracking seismic progress.

With Sinclair as a strong financial partner, IIAPCO planned to expand and speed up its seismic program. It was decided I would continue to run the Jakarta operation until March 1968, while Sinclair phased in their own management team. I would remain until June to be sure the transfer went smoothly. I was automatically excluded from continuing as resident manager, because it was Sinclair's policy not to allow a manager to own an interest in a project.

It was apparent that our relationship with the U.S. Embassy was improving now that Sinclair, an accepted, large American corporation, was IIAPCO's partner. It was, however, a disappointment to realize that "big" and "small" companies differed in importance in the eyes of U.S. government officials.

I continued to be frustrated with the lack of creative time to do geology. There was so much that needed doing in preparation for IIAPCO's operational start-up and there were so many meetings with Permina and other government agencies that my time for the drafting table never seemed to arrive. After several months of frustration, I relaxed in the knowledge that someone had to direct these functions and it was a responsibility I was not ready to relinquish. IIAPCO's relationship with Permina, and the Indonesians in general, was second only to that of Caltex. Maintaining that relationship was a full-time effort.

Pan-American's manager during this era had little understanding of Indonesia or Indonesians. He spent a great deal of time jogging and playing tennis with the U.S. Ambassador. Pan-Am did not have the insight to correctly analyze the Ibnu-Bratanata struggle. IIAPCO's excellent relationship with Permina and the Indonesians can be attributed to the fact that only one expatriate and his family represented the company. I had the authority to be decisive. I had a substantial financial interest in the project. It had to work. I had to learn and know the Indonesians and to see the Indonesian viewpoint.

Later, when IIAPCO set up an office to run its second contract area, we formulated an expatriate expense account policy that promoted at-home entertaining of Indonesians and their families, while restricting entertainment of other company expatriate personnel. Our endeavor to instill this philosophy in Sinclair met with only limited success.

IIAPCO's current primary concern was to get the seismic program under way. United Geophysical would service their boat out of Singapore, 1000 kilometers (600 miles) away. Our initial seismic program would be for 3200 kilometers (2000 miles) of air-gun, plus some sparker. Ray Halsey and Luther Horton, both Sinclair geophysicists, arrived for the start-up of operations. Permina set up a team to help IIAPCO obtain all necessary permits. These endeavors were as new to them as to us. Seismic data collection commenced July 26, 1967.

Larry Barker was handling the gradual turnover of IIAPCO's operations to Sinclair. As the summer progressed, Sinclair personnel came and went in Denver and Jakarta. Up to this point, my informal correspondence with the original IIAPCO partners discussed both business and personal matters. It included the humor and color of our Indonesian efforts. Now we were becoming formal. Correspondence followed an outline of only business. We sent weekly letters, month-end summaries, and quarterly reviews. I even began to send a detailed weekly letter to Permina. We had developed operating procedures.

Machmud and I were trying to obtain Presidium approval for Sinclair, Natomas, Carver-Dodge, and Warrior International as partners of IIAPCO. Dr. Sanger could not understand why we needed this. He thought that somehow it would alter the contract. Some Indonesian newspapers reported that IIAPCO had sold out to Sinclair. *Kami*, a Jakarta daily, wrote,

> Permina signed a contract for exploration and exploitation in the offshore of the Java Sea with IIAPCO, an American oil company, who appeared to have sold all its shares to Sinclair Oil Co. We therefore very much doubt the bonafidity of IIAPCO, who seems to be a company just emerged as a broker and which before going into exploration sold all it's [sic] shares.

We spent a great deal of time and effort trying to explain Sinclair's participation. Ibnu understood it. Kwa understood it. No one else seemed to.

In early August 1967, I stopped by Permina's office to inform Colonel Pattiasina that we had just sent the first seismic tapes to Houston for processing. He was extremely upset that I had done this without first getting his approval. I then went to see Dr. Sanger, who was acting as coordinator. He was upset with me for seeing Colonel Pattiasina when I should be channeling through his department. Sometimes it was difficult to win. Everybody thought they should be first in line, except for Kwa, who just grinned at my protocol problems. I still had been unable to find a good secretary capable of typing English.

The brute-stack of our earliest seismic data indicated a thick sedimentary section, substantial rollover, and no evidence of intrusives. An early line extended north beyond our contract boundary. As anticipated, the section thinned onto the Karimundjawa Islands, but surprisingly, after crossing the island chain, it again thickened. I immediately proposed to Permina an extension to our contract area to the north and west. Competition was becoming keen for new areas. We had no time to lose.

I was finding too few hours in a day to get the program going; be tour director for Sinclair, United, and Raydist; work out a Permina-requested "Detailed Arrangement;" prepare enough geologic data to keep the draftsman busy; properly organize accounting, past and future; work on a new contract proposal; make plans for future expansion; trot back and forth to Singapore, and keep everyone happy. In Denver, Larry was starting to investigate drilling rigs. We were hopeful that Sinclair-IIAPCO could spud its first well within a year.

Dave Dodge came to Jakarta for three weeks to help. We joined Horton for two days on the seismic boat in late August. We reviewed other areas of potential interest, including Ceram, Irian Jaya, and some hard-rock prospects in addition to the proposed extension area. We had a Shell geologic report on Irian Jaya, which represented data accumulated over a 30-year period. As it was a very negative report, we did not follow it up. Later, in 1971, Natomas, who by then had their own staff geologist to second-guess Larry, me, and all of IIAPCO, would not permit us to join the Irian Jaya venture put together by Trend Exploration.

On August 31, 1967, Dave and I met General Ibnu to express our interest in the adjoining area. Ibnu indicated he would give preference to IIAPCO, if our bid matched or bettered Union's bid. We presented Kwa a hurried draft contract on September 5. We immediately obtained U.S. approval from IIAPCO, but not from Sinclair. Roger and I made a few requested changes, and I presented a revised contract for Offshore Southeast Sumatra to General Ibnu on September 8, hopeful of a signature. However, he informed us it would be presented to the "oil committee" for comparison with other proposals. Ibnu felt this would avoid political criticism.

As Union was seeking three areas, including Offshore Southeast Sumatra, and as we were told IIAPCO's proposal for it was better, we felt we had the upper hand. The meeting with Ibnu had prompted him to get Presidium approval of IIAPCO's partners off center. On September 7, all of IIAPCO's partners were approved by General Suharto. Now we could relax!

Although we had no mutual area of interest, IIAPCO felt all of its partners should have the opportunity to participate in any adjoining areas. Sinclair held the determining vote on Offshore Northwest Java, a vote the IIAPCO group agreed to because of Sinclair's disproportionate early expenditure requirement. This would not be the case on a new area. Still, Sinclair wanted the controlling vote. As the IIAPCO group refused to give them more votes than ownership, an AMI (Area of Mutual Interest) agreement was never signed.

I had become good friends with a number of people from the Japanese oil and trading companies and I exchanged data with them. Piecing together information from Kyushu's seismic with IIAPCO's gave us additional evidence that a basin existed to the north of our contact area.

After a golf game with Ibnu, Andrew Alexeiev told me the General indicated he was going to grant IIAPCO the extension area. I was seeking New York and Denver approval for permission to move fast to obtain the area. We had much more data than the others, but we would soon need to give it to Permina.

The brute-stack seismic data indicated numerous large structures. Dick Brewer, a geophysical consultant, reviewed the early seismic and reported that it had been a long time since he'd seen such excellent structural indications at such shallow depths. He thought we had a winner.

Natomas had a 5% direct interest in the contract and had recently purchased a small stock interest in IIAPCO. Encouraged by the early seismic, Natomas desired to increase their participation even more. Expenditures were increasing rather rapidly as our work continued. Plans were under way to expand the program. In June 1968, we planned to start a $3,000,000 drilling program. My pocket was empty and Larry had a big hole in his. Pat just ignored his pocket. Larry opened merger discussions with Natomas.

IIAPCO's corporate interest after Sinclair's joining was 31.85%. IIAPCO's share of expenditure obligations was:

Expenditure	IIAPCO's share
1st $500,000	$32,000
2nd $1,000,000	$96,000
3rd $1,5000,000	$192,000
4th $2,000,000	$320,000
Thereafter,	31.85%

We had expanded our first

year's expenditure far beyond our $300,000 commitment. Sometimes New York and Denver neglected to inform me of expenditure increases and, consequently, I could not advise Permina, which wanted to be treated as a partner in these decisions.

In mid-September, I finally hired a secretary and Nancy was able to turn more of her attention to family matters. Ben Samsu, a young Permina geologist, was assigned to help IIAPCO in the implementation of our contract. He was also helping us gather onshore geologic data, straightening out formation names, and helping to find and construct type stratigraphic sections. We were accumulating data on other producing areas, particularly Sumatra.

It was most difficult for Larry to turn the reins over to Sinclair. Larry and some of the others of IIAPCO's original group wanted to help show Sinclair the way. Certainly IIAPCO's deep involvement in the project over a long period gave us an edge. We started to bruise the egos of some within Sinclair. As I had substantial ownership in IIAPCO, and was now under a one-year Sinclair employment contract, I became their best conduit to present drilling, seismic, and geologic information without agitating the operator.

Information among and between the participants was functioning, albeit with a bit too much duplication. Ed Chittick, who was responsible for running the Sinclair-IIAPCO operation from New York, arrived to become acquainted with the area, the people, and the problems. Plans were being made to do an additional 2400 kilometers (1500 miles) of seismic in 1967.

Machmud was busy with Permina, reviewing Indonesian law that might relate to our operations, as well as ramrodding the new survey crews. We were busy working on the 1968 budget. It had become apparent that I should stay away from Permina's head office, except when I had a definite purpose. By showing my face frequently, it only increased our operational problems as it allowed everyone to add his two cents. On October 31, the bids for the Offshore Southeast Sumatra area were closed. (Later they were reopened.) IIAPCO offered a $500,000 signature bonus. Bids were also submitted by Pan-Am, Union, Frontier, Hunt, and Southeast Asia Oil and Gas.

Bratanata was finally dismissed as Minister of Mining in late October, 1967. The newspapers reported that several large companies negotiating through Bratanata were reconsidering their plans of investment. The Presidium was abolished. Sanusi, who had been a prominent Presidium member and had been something of a problem for IIAPCO, became Minister of Education and Culture, at a safe distance from the oil business.

We revised our 1967 budget upward, from the committed $300,000 to $1,500,000. Conoco signed an onshore Production Sharing Contract with a $1,000,000 signature bonus. The entry price was climbing. Sinclair ignored Mr. Kwa as he passed through New York on his first trip to the U.S. Our biggest concern in turning over operations to Sinclair was that they be sure to maintain the excellent Permina relationship IIAPCO had worked so hard to build. I was trying to build a close operator/non-operator relationship. This was not an easy task, since Sinclair was not ready to listen to novices, even though we were offering hard-earned advice.

Our initial seismic grid was laid out in a 10x20-kilometer north–south pattern in the eastern half of the contract area. Some refraction work was done that indicated 2300 meters (7500 feet) of sediment in the troughs' axes. New York was concerned that we were not seeing evidence of the 4500-9000 kilometers (15,000-30,000 feet) of sediment seen in the troughs of South Sumatra.

Crewman to captain: "That water is certainly clear—just like the barrier reef off Australia." Bump! The United seismic boat had had two rudders, but now had only one! There were more delays while the boat went to Singapore for repairs, but on the way, we shot 240 kilometers (150 miles) of seismic across the western end of the area we were seeking.

In the first three months, we completed only 2400 kilometers (1500 miles) of seismic. Due to equipment breakdowns, we were six weeks behind schedule.

In November 1967, I travelled to Tokyo to meet Larry and Ben Tyran to discuss trading my IIAPCO stock for Natomas shares. I knew I must do something soon, before I went broke getting rich. Larry and I agreed to trade our controlling IIAPCO interest for Natomas stock. Pat thought Larry and I were trying to outmaneuver him and was opposed to a Natomas merger.

Pat and Chesley Pruitt played hard to get and delayed the merger until January 1968. The IIAPCO trade amounted to $1,500,000 of stock at Natomas's price at the time of $18 per share. If we established commercial production within two years, we were to receive an equal amount of contingency shares valuing IIAPCO's total trade at approximately $3,000,000. At an IIAPCO meeting, shortly after production was established in 1969, Chesley remarked that we should have traded for $10,000,000, not $3,000,000. When I asked what the difference was between three million and ten million dollars, he looked me straight in the eye and replied "Seven million dollars!"

Now that I have more than two nickels to rub together, I agree with Chesley, of course. Long after I left Natomas, he called to tell me he wanted to eat crow, saying we could not have picked a better company with which to merge. Natomas was selling at $18 per share and had a market value of approximately $65,000,000 at the time of IIAPCO's 1968 merger. After discovery in 1969, Natomas stock hit a high of $130.50 per share. In 1983, Diamond Shamrock paid $2,350,000,000 for Natomas, including the assumption of $850,000,000 of long-term debt.

In late 1967, we were still unable to communicate with the seismic boat by radio because of interference from amateur transmitters. Compilation of onshore geologic data, which was needed to better interpret our seismic, was progressing. We were finally getting a good land survey to tie in our seismic.

When Professor Soemantri was appointed the new Minister of Mining, Ibnu asked that the Director General be put back under control of the Ministry. We had a steady stream of supply and drilling companies calling on the Jakarta office. In New York, Sinclair was becoming more involved. Contracts were signed for an oceanographic study and a bottom coring program. Drilling bid requests went out. ADC, a Canadian company, signed a Production Sharing Contract in which Cities Service would participate and operate.

On December 8, 1967, Colonel Judo Sumbono asked me to go to Singapore to call Ibnu in Tokyo. He and Kwa wanted Ibnu to know that strong rumors coming from sources close to Suharto indicated he was being replaced. They felt their Jakarta absence might create a problem. I became their emissary and called Ibnu with this information. After a short silence Ibnu thanked me and in effect said, "So what else is new?" Somehow Ibnu came out on top once more.

Sinclair was as unhappy to learn of my involvement in this political affair as I had known they would be. In late December, I received a strong letter from Chittick informing me that the operation would be run Sinclair's way and indicating that independent thought and action would not be tolerated. He also had a few words to say about Larry Barker. We obviously didn't fit Sinclair's format and it didn't appear that we ever would. We had encountered few, if any, problems with management throughout negotiations. However, once Elston and Chittick were assigned to the operation, we began to have conflicts. This was expected, but disappointing. Our motives were suspect, and in some cases Sinclair personnel resented our being there.

By mid-December we had shot a total of 7150 kilometers (4430 miles) of seismic. We were now averaging 100 kilometers (63 miles) of seismic line per day. Pete Petersz, of Sinclair's Brussels office, came on temporary assignment; the beginning of Sinclair's phase-in as operator. Sinclair was anxious to take over my desk. I had no objection, except that I wanted to look over their shoulder for awhile. I didn't want Sinclair to mess up IIAPCO's good relationships.

Against my advice, in early January 1968, a document was drawn between Permina and Sinclair, rather than between Permina and IIAPCO. Dr. Sanger returned it with a letter stating, "Up to now, IIAPCO is Permina's exclusive partner with regard to Offshore Northwest Java." Sinclair was so anxious to have their name on the door they wouldn't listen. The transition needed time.

IIAPCO's bid for Offshore Southeast Sumatra was not the best offer. Southeast Asia Oil and Gas (who was a Mr. Brinsmade) had offered a $5,000,000 signature bonus. Ibnu told me that he was quite sure Brinsmade did not have the money and that he did not expect the proposal to close. He said, however, that it would be politically necessary to give Mr. Brinsmade time.

At a New York meeting in January 1968, I indicated that I felt Ibnu would give the area to IIAPCO, if Brinsmade failed to raise his money. I thought it would require a $2.5 million signature bonus and I asked for flexibility in a new proposal. The IIAPCO Group gave me an okay, but Sinclair refused. They would not give me any open authority. After returning to Jakarta, I submitted a revised proposal to Permina, which included a $2,500,000 bonus. We needed to move fast. The IIAPCO Group was in accord and, although Sinclair had not yet agreed, Larry and I felt they would go along.

When I.D. Stevens arrived in February to replace me as Sinclair's Operations Manager, I took him to meet General Ibnu. On Chittick's instructions, he proceeded to inform Ibnu that Sinclair was not a party to IIAPCO's negotiations for the extension area and that Sinclair's financial and logistical support would not be available to IIAPCO for this area. That information didn't concern Ibnu in the least. As Roger and I were on Sinclair's payroll, we were no longer allowed to be involved in formal negotiations, although we continued to discuss the area with Permina. Larry took over as IIAPCO's formal representative.

By mid-February, 1968, we had completed the reconnaissance seismic in the eastern half of our contract area. Sinclair's John Grosso arrived to oversee the sea-bottom coring and John Cerni came to set up and supervise accounting. Barker was there. We were still using Nancy for overflow typing. Permina signed its ninth Production Sharing Contract with Union. General Hartawan was named Chairman of the Coordinating Committee for Offshore Contracts (DKKA). In early April, we met with Permina Unit III to review IIAPCO's progress and to formally turn my duties over to Sinclair. We had completed 8500 kilometers (5300 miles) of seismic. The office was a zoo. We were fast running out of space.

Southeast Asia Oil and Gas failed to raise the bonus Brinsmade had

The Drill Ship E.W. Thornton, drilling on the PSI-A-1, September 1968.

offered for Offshore Southeast Sumatra by the new March 31 closing date. Ibnu indicated IIAPCO would be next in line for discussions. Other parties, however, were now attempting to upgrade their proposals. Ibnu decided to extend Brinsmade's deadline until June 1 to avoid political criticism. He did not, however, consider the late upgraded proposals. He told me one major had offered him a personal bribe to allow them to resubmit a better proposal.

Elston and Chittick began to relax as I endeavored to back away from actively participating in management. I continued to observe and push for more interaction between Sinclair and Permina. As Stevens assumed Jakarta management, I found time to visit Caltex's Minas operation in Central Sumatra.

Minas is the largest oil field in Southeast Asia and one of the largest in the world. The drill site was staked by Caltex, but World War II intervened before it was spudded. The discovery was then drilled by the Japanese, the only wildcat drilled by them during the war. [In the early 1980s, I visited the Japanese geologist who sat this well.]

Kyushu invited me to visit Maselembo, an island halfway between Madura and Kalimantan, where they were building a supply base. The islanders watched with fascination as the ship unloaded a truck and bulldozer. These secluded, simple-living people had never seen a motorized vehicle.

By early May, the reconnaissance seismic had essentially been completed in the western half of our contract area and 2900 kilometers (1800 miles) of detail line was begun. The structures in the western half were larger, had greater relief, and a thicker section than those in the eastern half. One anomaly exhibited 90 meters (300 feet) of closure and covered 285 square kilometers (70,000 acres). Permina approved Reading and Bates as our drilling contractor. We would use the catamaran drill ship, *E.W. Thorton*.

Mr. Steiniger retired as Chairman and CEO of Sinclair in May 1968. He had been an important factor in Sinclair's decision to join IIAPCO. Machmud was told there was no future for him with Sinclair outside of Indonesia. Chittick wanted to be sure IIAPCO had not brainwashed Machmud before he decided whether Sinclair would keep Machmud on their payroll. Ultimately they did keep him and he became an important factor in Arco's Indonesian presence.

It was announced that Pertamin, the other state oil enterprise, would be merged with Permina on July 1. General Ibnu Sutowo was to be appointed President Director of "Pertamina," the merged company. Pertamina's Board of Directors would be appointed by the Minister of Mining. General Suharto had recently been elected President of Indonesia.

After final calls on friends at Permina, the ministry and the U.S. Embassy, we were gone. Our two Denver daughters joined us in Bangkok. We forgot about Indonesia as we toured Asia and Europe for two months on our way to our new home and office in Denver.

It was my time to leave, for many reasons. Pertamina, the DKKA, and Jakarta were becoming organized, as was the Sinclair-IIAPCO operation. I had no place in an organized environment, as necessary as that environment was, be it corporate or government. We were sorry to leave, but glad to go.

IIAPCO'S SECOND CONTRACT

Ibnu gave Brinsmade a third extension, this time to July 1. Barker made trips in June, July, and August to Jakarta in an effort to obtain the new area for IIAPCO. Foreign banks were being allowed to open Jakarta branches. Indonesia had signed 50 agreements in various ventures, including oil.

By August, I was in Denver. Sinclair was now aware of the value of the new area IIAPCO was seeking. Unfortunately, Sinclair had shot themselves in the foot, cutting themselves out of what was to become a successful second venture. After their February performance, IIAPCO wasn't about to let them back in. In August, Bunker Hunt tried to convince Barker that Hunt's chances at the contract were equally as good as IIAPCO's. Hunt claimed to have a high-ranking and influential military person behind them. They wanted IIAPCO to cross-assign rights, so that no matter who won, the other would end up with half the deal. Barker called me in Denver to discuss Hunt's proposal. Ibnu told me before I left Jakarta that IIAPCO was next in line. I did not believe Hunt's "outside influence" would have any bearing on the awarding of the contract. Barker agreed, and we went for broke.

IIAPCO was awarded the Offshore Southeast Sumatra contract on September 6, 1968. It was signed by General Ibnu on behalf of Pertamina, by Larry Barker for IIAPCO, and was approved by Professor Soemantri, Minister of Mining. This second Production Sharing Contract covered 130,000 square kilometers (32,000,000, acres). We paid a $1,250,000 signature bonus and an additional $1,250,000 six months later. IIAPCO agreed to spend $22.5 million in the first ten years. Natomas's stock was now trading at $35.50, up from $18 at the time of the IIAPCO merger. IIAPCO, now a Natomas subsidiary, had a 68% interest in this new Offshore Southeast Sumatra contract. The remainder was owned by Carver-Dodge and Warrior.

EXPLORATION AND DISCOVERY

Larry and I crossed paths in Singapore, he on his way home with the new contract and I on my way in to join the Pertamina-Sinclair-IIAPCO (PSI) spud-in ceremony on September 7, 1968. Carver, Tyran,

Joe Kennedy (a Sinclair V.P.), I.D. Stevens, and I were received at the Palace by President Suharto as part of this ceremony. Joe Obering joined me on the drill ship. The slowest drilling time on this first test was 18 meters (60 feet) per hour. I returned to Denver to help with exploration plans and a Work Program and Budget for the new area.

Continental was awarded a second contract in the South China Sea for which they had bid a $7,000,000 bonus. IIAPCO had indeed been fortunate to be first. Oil was still only $1.62 per barrel.

Sinclair abandoned the first test on the "A" structure as a dry hole at 1163 meters (3816 feet). Our second test, on the "B" structure, encountered porous sands with good cut, but was abandoned after testing some gas. This feature later proved to be a good field. IIAPCO and the non-operators were having difficulty obtaining basic geologic data from Sinclair, even though the non-operating group was now paying 54% of all bills. The IIAPCO-Sinclair Indonesian venture took a back seat as Sinclair and Arco began merger talks.

Bob Elston had an inherent distrust for the Indonesians, an attitude reflected by Chittick and to some degree by Stevens. Sinclair refused to go out of their way or make any effort to see or entertain Pertamina personnel visiting the U.S. IIAPCO bent over backwards to do so. Sinclair now had eleven expatriates plus a number of nationals working out of the small IIAPCO offices. The entire house was being converted into office space.

On November 3, 1968, an article in the *London Sunday Times* quoted one major company's reason for being reluctant to join the Indonesian oil rush: "If we accepted the terms here, we'd make all other countries where we have fields want to cancel their contracts for new ones. " The article also cited "Pertamina's insistence on joint management," as further reason for hesitation.

Sinclair spudded our fifth test, the E-1, on December 11, 1968. At a meeting in New York, in mid-January 1969, Sinclair disclosed that on December 20 the PSI E-1 had had a blowout and flow of a substantial quantity of fluid, including at least 35 barrels of oil. The well was out of control for 20 minutes with mud, gas, oil, and fist-sized chunks of rock blown halfway up the derrick. Sinclair had previously reported this as a "small show of live oil" with "a few beakers" to the surface.

I analyzed all the data available, including sample descriptions, gas recorder, and engineering data along with IIAPCO's independent log evaluations. I sent a memo to the non-operating IIAPCO partners, concluding that I believed the well was capable of flowing in excess of 2000 barrels of oil per day (BOPD). Ben Tyran sent my memo on to Sinclair's Bob Elston, who was not at all happy with my conclusions.

On January 18, Sinclair reported that the E-1 tested at the rate of 360 BOPD from 1043–1049 meters (3421-3440 feet). When I told a Phillips geologist about the test, he said it would never make it and informed me he had studied Indonesia for two years and concluded our area had little potential.

At Elston's request, Sinclair's Manager of Exploration in New York critiqued my analysis of the E-1, calling it a "commendable exercise." Besides putting in my two cents, Barker had been critical of Sinclair's operation in general. Elston thought his IIAPCO contact should only be through Natomas's Ben Tyran. He informed Ben that Larry and I were endeavoring to second-guess them and were wasting their time. We thought they were wasting our money. Ben didn't know whom to believe.

It is indeed difficult to do a job when one's own top management vacillates and is not supportive. It kills creativity. I began to have more and more problems with Ben and, on occasion, with Chandler Ide, President of Natomas. I had been self-employed for my entire career, and I thought, acted, and worked independently. It was most difficult for me to fit into a corporate mold. At Natomas's request, Larry had moved IIAPCO's corporate office from Denver to San Francisco. I kept my IIAPCO office in Denver, because I knew I couldn't work near Tyran.

On February 25, 1969, an article in the *Wall Street Journal* reported a "significant" Sinclair Indonesian oil discovery. The PSI E-1, on a stabilized test, flowed at the rate of 2,600 barrels of 36° API oil through a 1-inch choke from 2230 to 2260 feet (680 to 689 meters). The article stated, "The discovery could prove one of the most significant to date in Sinclair's effort to become more of an international company." Sinclair's stock closed up $7 at $105.50, Natomas closed up $9 at 43^{5}/_{8}$.

The producing zone was one of the zones deemed productive by IIAPCO's log consultant and referred to in my much-criticized January memo. Further tests indicated the combined rate from the four tested zones totaled 4650 BOPD and 38 MMCFGPD. Three additional zones had potential. I was vindicated. When I briefed General Ibnu on the results of the well, he grinned and told me, "Everybody here has always considered you as a bit of a foolish optimist. Now we think you are a very clever fellow."

Sinclair thought any discovery drilled from the drill ship should be treated as a disposable well. We thought they should be only temporarily abandoned. For IIAPCO to qualify for the $9,000,000 Sinclair production loan, it needed 30 days of production averaging 3000 barrels of oil. This would pose a problem if we continued to drill disposable wells. Long-term testing from the drill ship was not feasible.

I rented space and opened the second Jakarta office to handle IIAPCO's new Offshore Southeast Sumatra operation. We commenced a 4800-kilometer seismic program in February, using new satellite and sonar Doppler equipment for positioning. Never again would we have to worry about our German surveyor. Larry and I alternated with frequent Indonesian trips, while we sought an expatriate manager and a geologist for Jakarta.

Sinclair merged with Arco on March 3, 1969. Arco, their hands full and their budget committed to their new North Slope discovery, had a new stepchild on the "South Slope." It was some time before Indonesia was given serious Arco budget consideration. Arco believed it would not be practical to consider a start-up of production before late 1970, when development on the "E" structure was expected to be nearly complete. The non-operators felt that an early start-up would politically benefit Pertamina and Indonesia, boost Indonesian morale, and start our needed cash flow.

In May 1969, Caltex produced its billionth barrel of oil from the Minas Field in Central Sumatra. Japex tested 5500 BOPD on an offshore North Sumatra wildcat. IIAPCO had completed its 4800-kilometer initial seismic program on the new area. Preliminary interpretations indicated 5 to 10 drillable prospects.

I had been in Singapore in January 1969 with George Temple, executive vice president of Reading and Bates, when we learned of the first successful "E" test. Reading and Bates wanted to get a piece of the action, so I suggested they talk with Carver-Dodge (Cardo). In May 1969, Cardo traded their 12.25% interest in Offshore Northwest Java and 22% interest in Offshore Southeast Sumatra for $25,000,000 of Reading and Bates stock.

In the Arco area, the PSI E-2, 1.8 miles to the northwest, successfully confirmed the "E" discovery. On May 14th, Natomas stock closed at $96^5/_8$, rising to $130^1/_2$ in a down market, before dropping off in mid-June. I was busy hunting for houses and looking for a larger Jakarta office. Ed Englehart joined IIAPCO as Jakarta manager, and R. C. Slocum joined as our geologist. They were to be responsible for the Offshore Southeast Sumatra area and keep an eye on Arco's operation. P. C. Smith, a petroleum engineer, and Ed Reid, a geologist, joined IIAPCO's San Francisco staff. These four remained IIAPCO's top team until Larry and I resigned from IIAPCO in 1971. Englehart and Reid, along with nine other former IIAPCO hands, now work with Barker in Pan Pacific Petroleum.

By the fall of 1969, as IIAPCO's Sinclair relationship improved, our relationship with Natomas's Ben Tyran deteriorated. Neither Ben nor Chandler Ide understood oil exploration or exploitation, yet Ben wanted his fingers in everything IIAPCO did and Chandler didn't stop him from interfering. We were not even allowed to determine our own staff policies. If we felt a draftsman needed and deserved a raise, we had to clear it with Natomas.

I was now constantly running between Denver, Jakarta, San Francisco, and New York. By December 1969, Pertamina had signed 30 Production Sharing Contracts. The Indonesian Petroleum Club opened with 77 charter members.

After a year-and-a-half of calm waters, Ibnu once more came under fire, this time in a series of newspaper articles in the *Indonesia Raya*. These began in November 1969 and continued into 1970. Even I made the headlines on December 15 with:

PERTAMINA IS SELLING THE ASSETS OF THE REPUBLIC OF INDONESIA
Donald F. Todd, Okada, Harold Hutton, Morris PalmerUsed by Ibnu Sutowo as Middle Men in Order to Facilitate Manipulation of Grand-Scale Commission Fees

In referring to me, the translated article stated in part:

> Ibnu Sutowo granted him an offshore mining area on the northern part of Java with an acreage of 5.4 million hectares…. IIAPCO is a company intentionally founded by a group of Americans in Denver, Colorado, specifically to accommodate the "gift" from Ibnu Sutowo. One should not be surprised if its financial ability is doubted by executives of Bank of Indonesia at that time. The doubt proved to be confirmed by the fact that IIAPCO sold 51 percent of its interest to Sinclair Oil (US) after it got the contract in its pocket. Sinclair Oil is, indeed, capable to exploit the oil mining area presented by Ibnu Sutowo to Donald F. Todd. Why was said mining area not directly awarded by the RI Government to those who are capable to exploit it?…. Amidst the downturn in the NY Stock Exchange, some exceptions are obvious. The highest position among the shares is still maintained by Natomas…which is active in the "Indonesian Oil Play" … because of this play, the shares of Natomas are skyrocketing…. It is obvious that in only seven months time the value of the shares increased by 139% or an average of almost 20% per month. Against the lowest figure in 1968 an increase is registered of 272%…. It is really a golden opportunity to become

very rich quickly and without any effort.... Through the holding company are traded the shares of the venture discovering oil offshore Java. For example, a guy like Donald F. Todd, who owns a considerable number of said shares, stays "relaxed" while enjoying the news about increases in the value of his shares. Anyhow, it is another party who bears the risk in said venture. However, "key-money" has been received while his capital is "nil"...with the blessings from Indonesian officials only.

The owner and editor of the *Indonesia Raya* and the author of most of the articles was Mochtar Lubis, an Indonesian Ralph Nader. He was a respected journalist writing anti-Ibnu stories under the flag of a "free press." He obviously spent much time collecting data and was ruthless in his misinterpretation and misuse of them.

Indonesia was an emerging nation, and the Western world was endeavoring to force Western culture upon it. To the author of these articles, a General Motors, a Ford, or a Sinclair would represent ability, intelligence, and money. Lubis never stopped to consider that Henry Ford built one automobile before he built a thousand.

It was difficult for most Indonesians to understand our free enterprise system, and our ways of conducting business looked downright sinful to them. It was a sin to create a company to obtain the contract. It was a sin to bring in monied partners. It was a sin that Natomas's stock had increased in value just because oil was discovered. This aptly reflects the "corruption" problem so frequently faced by Ibnu. Ibnu, although he was very much an Indonesian, was quite comfortable with the free enterprise system and equally at home in the West and in the East.

The dramatic and rapid growth of Pertamina and Indonesia's oil production and income in the post-Sukarno era was largely due to the determination of this strong-willed, autocratic, intelligent individual. Ibnu's ability to make a quick decision and to get a job completed, his red-tape-cutting policies, his individualism, his integrity, and his self-assured, cocky nature had made Pertamina the most efficient and successful Indonesian state enterprise, and at the same time had brought Ibnu great personal criticism and accusations of corruption. An article in the *Far Eastern Economic Review* of January 1, 1970, stated, "Those who fault Sutowo should not ignore Pertamina's grand record."

In November 1969, Pertamina tested 2000 BOPD from Oligocene andesites at Jatabarang, onshore from our Northwest Java area. The test was taken after a gas-kick was recorded on the gas unit Larry and I had given Permina. I had used this portable unit on our Montana wells. Arco suspended drilling on the "B" and "E" structures, pending

An Indonesian Experience 107

The Cinta platform, September 1970.

the arrival of a development jack-up rig. In December, the "F" structure tested oil. Arco and IIAPCO were enjoying a magnificent discovery rate on the Java block. Ten wells had been drilled in the offshore northwest Java area. Four were commercial, one was noncommercial, and five were dry.

In December, C. Itoh, a large Japanese trading company, agreed to purchase from Natomas a 7% stock interest in IIAPCO, for $21 million, giving us a little breather in this costly venture. In addition to this sale to C. Itoh, Natomas financed this fast-developing play through the sale of its subsidiary, Pacific Far East Lines; a new Natomas public stock issue; a loan-joint venture agreement with Shell; and bank borrowings.

By the end of 1969, Shell had reentered the Indonesian oil play, Mobil had begun a North Sumatra exploration program, and Caltex had stepped up Minas production and had an active exploration program underway in Central Sumatra. In South Sumatra, Stanvac had rejuvenated their exploratory efforts. Union, Cities Service, Kyushu, Japex, Continental, Phillips, Agip, and others had seismic work in progress. Indonesian crude, with its low sulphur content, was beginning to command slightly higher prices. Japanese imports of Indonesian crude increased 68% from the previous year. Most of the attractive offshore areas were now under contract. Pertamina's earnings were providing 48% of the Indonesian government's total revenues.

In the Offshore Southeast Sumatra area, IIAPCO had averaged 240 kilometers (150 miles) of seismic per day and had completed 14,800 kilometers (9200 miles). The first test of a ten-well exploratory program in this new area spudded in January 1970, using a jack-up rig. These tests were positioned by satellite, an industry first. IIAPCO abandoned the first test in the Sumatra block as a dry hole. The second well tested encouraging shows, but was also abandoned.

In June 1970, *U.S. News and World Report* stated, "A major problem under Suharto, as under Sukarno, is still graft.... The oil business is riddled with corruption because it is obviously a rich source of Indonesian income. Producing companies are squeezed at every turn, mainly by the Army." IIAPCO and Arco certainly weren't being squeezed. This reporter appeared to be doing his creative writing on a bar stool.

IIAPCO's Sumatra prospects were named after the daughters of Pertamina friends. IIAPCO spudded Cinta #1, the ninth test in the area, on August 10, 1970. Proposed as the third test, it was postponed so an additional 10 feet could be added to the jack-up legs. Cinta #1 reached a total depth of 1146 meters (3760 feet) on August 26. Substantial oil flow rates were tested from three separate zones, prompting IIAPCO to install Indonesia's first production platform.

OFFSHORE PRODUCTION COMMENCES

Before IIAPCO began drilling in the second area, we had designed and built a one-well completion platform that could be floated to location on short notice. In September 1970, after successfully testing the well, this platform was installed over the Cinta #1. I arrived in early October to organize a dedication ceremony. I met with General Ibnu and suggested he invite President Suharto to join the ceremony. Ten days later, on October 23, 1970, President Suharto dedicated the Cinta platform and opened the valve starting Indonesia's offshore production.

Pertamina had suggested to Arco, when they first had tested oil on the "E" structure 19 months earlier, that an early start-up of production would do a great deal for Indonesia's morale and be politically beneficial to President Suharto and Pertamina, as well as to Ibnu. Suharto needed, and the public wanted, an "accomplishment." IIAPCO had tried to convince Arco to do this. Arco wanted to have at least one platform having a capability of producing 25,000 BOPD on sustained production before start-up. They also wanted to have a large storage vessel and mooring system in place. Arco could have had the honor of producing and shipping Indonesia's first offshore crude, had they not been so short-sighted and business-oriented.

President Suharto, General Ibnu, Mining Minister Soemantri, U.S. Ambassador Frank Galbraith, and other dignitaries attended this event. As Mrs. Suharto and Sally Ibnu were not joining the ceremony, Nancy, with Jane Slocum, planned only to welcome the President as his helicopter landed at a nearby island. Much to my surprise and delight, President Suharto invited Jane and Nancy to attend the ceremony as his guests.

The Sumatra contract had been signed on September 6, 1968 and the Cinta discovery was in August, 1970. In October 1970, less than two months after discovery and only 26 months after signing the contract, IIAPCO produced and shipped Indonesia's first offshore oil. In 15 days, testing on various choke sizes, we produced more than 50,000 barrels of oil. Our flow rate at the end of this period was 7728 BOPD. Our temporary storage was a barge and a 35-ton, inter-island tanker.

This makeshift, one-well platform was later attached to a permanent development platform. The Cinta #1 produced more than 1,000,000 barrels of oil from this "temporary" platform before being abandoned. Indonesia issued a postage stamp commemorating the Cinta platform and the start-up of Indonesia's offshore production.

By the end of 1970, foreign groups, including French, Italian, British, Dutch, Japanese, Australian, Canadian, and American interests, were exploring in 46 areas of Indonesia. New production had also been established by Union and Japex offshore eastern Kalimantan, by Asamera onshore in North Sumatra, and by Pertamina onshore in

President Suharto opening the valve starting Indonesia's offshore production.

western Java. Four new offshore and six new onshore fields had been discovered. Twenty-three seismic crews and 28 drilling rigs were active. For a change, Pertamina and Ibnu were receiving favorable press coverage. This press turnaround from one year earlier related to the many visible Pertamina accomplishments.

IIAPCO established production in the Kitty field in the Offshore Southeast Sumatra area in January 1971. IIAPCO now had 200 employees in Jakarta, San Francisco, and Denver. I was spending more and more time on personnel and administrative problems. I remained unwilling to move to San Francisco because I was still having difficulty with Ben Tyran. I'm sure that it was only my excellent relationship

The Cinta dedication, October 23, 1970.

with Mr. Davies, Natomas's chairman and major stockholder, that kept me on IIAPCO's payroll. Ben Tyran and Chandler Ide thought IIAPCO's executive vice president should reside in San Francisco. Since I refused to move, they were seeking someone else for this position. Davies created a new title for me, "Vice President at Large."

In February 1971, I reviewed Trend Exploration's program and proposal that sought partners onshore in Irian Jaya. Trend's contract covered the same area Dave Dodge and I had earlier turned down. I recommended that IIAPCO join Trend, but Natomas's geologist rejected it. I was so impressed with Trend's data and evaluation that Natomas's turn-down prompted my resignation from IIAPCO and Natomas in order to personally commit to a 10% interest.

I became Trend's first non-operating partner. They had had 60 previous rejections. Two weeks later, I talked Larry Barker into bankrolling me for another 10% and we formed Southern Cross Limited, with each of us holding half the stock. As Larry remained president of IIAPCO, he could only be an inactive Southern Cross shareholder. Larry asked Natomas to reconsider their turn down.

An excerpt from a second rejection letter by Natomas's staff geologist, Keith Davis, written in March 1971, reads:

> It is extremely difficult to prognosticate what amounts of production would actually be encountered. It is very easy to succumb

to the luxurious idea that production would be the same as in the Sirte Basin pinnacle reefs in Libya, which will produce in the 30,000 BOPD range. A moment's reflection would indicate that if a professional geologist gave the best wells (world-wide) as a target in his evaluations, then management would "buy" all plays—and would very probably soon be bankrupt.

And from the *Oil and Gas Journal* two years later:

> The Petromer Trend group has a fourth oil strike on its 1,275,000-acre (5,160-square-kilometer) contract area in Irian Jaya, Indonesia. The group's #1 North Kasim flowed at the rate of 31,680 BOPD of 27° gravity oil from 4,912–5,300 ft. (1,497–1,615 meters) in Tertiary carbonate reef.

This was the largest well ever tested in Southeast Asia. So much for corporate second-guessing!

In April 1971, we sold half of Southern Cross's interest to C. Itoh. As part of the trade, C. Itoh committed to carry Southern Cross's remaining 10% interest for its next $1,000,000 of expenditures. Larry resigned from IIAPCO six months after I did in order to become active in Southern Cross and to pursue other interests.

In September 1972, shortly before Trend spudded its first test, we sold 20% of our Southern Cross shares to Banque de Paris et des Pays-

From left: Don Todd, President Suharto, General Ibnu Sutowo, Minister of Mining Soemantri, and U.S. Ambassador Frank Galbraith.

Bas (Suisse). Trend had completed its initial seismic, which confirmed what we had believed to be reef photo anomalies. Trend's first test flowed 26,000 BOPD from a shallow Tertiary reef. Trend established production in six reefs before it became necessary for Southern Cross to dig into its own pocket to finance continued exploration and development.

From inception of production in 1973, through year end 1990, Trend had produced approximately 285,000,000 barrels of oil from this Irian Jaya Production Sharing Contract.

LIFE AFTER IIAPCO

At the time of my resignation from IIAPCO in February 1971, Arco had eight oil fields and three gas fields. IIAPCO had two oil fields in the second area. Arco's Ardjuna complex went on stream September 7, 1971, at 24,000 BOPD, which increased to 40,000 BOPD by December. IIAPCO began steady production of 50,000 BOPD from a permanent platform on September 10, 1971. In March, 1972, Arco dedicated its one million-barrel storage barge. By 1975, Indonesia's total oil sales accounted for 80% of the country's GNP.

One of Ibnu's key objectives in the Production Sharing Contract was to allow the expatriate oil companies to operate free of red tape. It was Pertamina's job to shield the companies from the mindless hindrances of government bureaucracy and to eliminate politics from our job effort. Ibnu's philosophy was to provide the expatriate companies with the tools and equipment necessary to do our job and to make available housing and other facilities of standards to which we were accustomed. He felt it was Pertamina's job to see that the expatriates had all the comforts of home, because we were there to help Indonesia become an aggressive oil producer.

Ibnu built a modern hospital for Pertamina and expatriate company employees. He not only built a fleet of inter-island tankers, but also an air service that comprised helicopters and prop and jet planes. Pertamina had barges and tankers available for offshore temporary storage so that operations might move rapidly ahead. Pertamina went into the insurance business and into communications. They had Indonesia's most sophisticated computerized accounting system.

Compared with other Indonesian state enterprises, Pertamina was so efficient and aggressive that when something needed doing in another field, President Suharto frequently asked Pertamina and Ibnu to front it. This put Pertamina into the steel business.

In 1972, Pertamina borrowed $350,000,000 without clearing it through the central Indonesian financial authorities. When the International

Monetary Fund learned of it in 1973, Indonesia lost its IMF stand-by rights, and U.S. aid was temporarily suspended. Pertamina's critics acknowledged its contributions and even admired Ibnu's effort to show Indonesia and the world what the country could do.

However, it was apparent to government officials that Ibnu's freewheeling style would have to cease. Pertamina was called the most efficient government department in Indonesia, but was also charged with running a state within a state. *The New York Times* reported on April 16, 1973, "Pertamina operates like a big foreign oil company, only more so."

By January 1974, the students were again rioting against "corruption." Other OPEC nations were criticizing Suharto and Ibnu for what they considered to be soft contract terms. As these IMF, OPEC, and student pressures increased, oil prices were escalating as a result of the 1973 Arab oil embargo. In Indonesia, crude prices had increased gradually from $1.21 per barrel to $3.80, from 1964 to mid-1973. In November 1973, crude had reached $6.00 per barrel, and by January 1974, the price had jumped to $10.80.

In December 1973, Senator Jackson first endeavored to introduce U.S. legislation imposing a "windfall profits tax" on domestic U.S. oil companies. The prevailing Congressional attitude was that such a tax was justified because the price increase was due to international politics rather than supply and demand. Indonesia used this excuse to impose their own windfall profits tax of 85% on crude values above $5.00 per barrel on January 18, 1974. They were merely following the U.S. example that "the oil companies sort of don't deserve it anyway." Even Ibnu—under pressure—subscribed to the Indonesian notion of "what the hell, the U.S. is going to get it, why shouldn't we?"

In December 1975, Arthur Young and Company was hired by the IMF and the Indonesian government to send an auditor to Jakarta for three years. In an interview, even before going to Jakarta, this individual indicated his mission was "to help straighten out the financial fiasco of Pertamina...." Pressure to control or replace General Ibnu came primarily from the IMF and the World Bank, and was supported by Suharto's team of technocrats.

Early in 1976, the foreign oil companies came under pressure to accept a change in the Production Sharing split. The U.S. Embassy was convinced a change would occur and that if the companies did not agree to a change, a government decree would be issued. They felt that once one company weakened, the game would be over. The embassy considered General Ibnu's departure only a matter of time. Tirto Utomo (formerly Mr. Kwa) thought the Indonesian government had their heads screwed on backwards. Ibnu refused to accept the government's position of changing the Production Sharing split.

The Indonesian government's stance was based on an IMF report, which went first to the technocrats and then directly to President Suharto. Pertamina was never consulted. Part of this report read, "Current oil company profits in Indonesia appear to be very high by historical standards and even more so by international comparison. Available data indicate that company profits per barrel in Indonesia is almost ten times higher than the average level in eight major OPEC countries."

This was grossly inaccurate. Indonesia, advised by the IMF, World Bank, and U.S. accounting firms, manipulated data to justify a contract change. The oil companies were up in arms. Only Trend, operating under the new Production Sharing Contract, was in a profit position and they had only been there for a few short months. Ibnu told Suharto to either give him room to move or to let him go. Suharto was upset with Ibnu over the whole affair.

EPILOG

General Ibnu was finally forced out of Pertamina in 1976. Shortly afterwards, Indonesia and Pertamina imposed an adjustment on all Production Sharing contracts, raising the 65/35 split to an 85/15 split in favor of Indonesia. This was in spite of the fact the contract barred annulment, amendment, or modification without mutual consent.

Ibnu never would have allowed this to happen. While he was "Mr. Oil" every contract became a bit tougher than the one before, but he never would have gone back on his word or his signature. Ibnu adamantly refused to consider this suggested governmental change, and I have always believed that his refusal was a major factor in his removal from office.

At the time that Indonesia was pressuring for changes, Ibnu asked me why the oil companies didn't stand firm and simply refuse. However, a domino reaction occurred after the first oil company gave in—which happened long after I had resigned from IIAPCO. Although money can still be made and is being made on Indonesia's Production Sharing Contract, I took this mandated change personally and began decreasing my involvement in Indonesia. I, like Ibnu, believed a deal was a deal.

In mid-1976, Piet Haryono replaced General Ibnu as president-director of Pertamina. Pertamina reportedly had a $10.5 billion debt and was near collapse. Ibnu was alleged to have been involved in overpricing and kickback schemes. Had he been a dishonest schemer, it is my opinion that the touch would have been put on me and IIAPCO long before.

I had signed the first Production Sharing Contract nearly 10 years

Don Todd with his *Pelepor* medal.

earlier, and never once did I have any indication or knowledge of graft or the desire of a kickback in any form, either from Ibnu or from any of his top assistants. Ibnu was eventually cleared of any wrongdoing and Indonesia owes a debt of gratitude to this individual—the man most responsible for the growth of the Indonesian oil industry over the past 25 years. Ibnu's autocratic style rapidly moved Indonesia back into the mainstream as a producer.

In 1974, Chandler Ide invited Larry and me to a party celebrating the 10th anniversary of the founding of IIAPCO. In a toast, Chandler said IIAPCO had found no significant new oil after the two of us left.

In 1978, I sold my Southern Cross shares to Larry Barker, ending 14 years of direct involvement in Indonesia's oil industry. My good friend Judo, now General Judo Sumbono, replaced Piet Haryono as Pertamina's president-director. Pertamina now periodically has a new president-director, but none can take General Ibnu's place.

Cumulative production through 1990 from the three Indonesian areas in which I was involved exceeded two billion barrels of oil and gas equivalent.

In October 1985, Pertamina and the Ministry of Mining invited Nancy and me to Jakarta to help Indonesia commemorate 100 years of petroleum development. The ceremonies were opened by President Suharto. At a dinner held October 8 for 2500 people, 16 honored guests were awarded *Pelopor* (Pioneer) gold medals by Minister of Mining and Energy Professor Subroto. Eight Indonesians, three Japanese, and five Americans received this award. My *Pelopor* award read:

> Based on his expertise as a geologist, in 1964 he was successful in convincing the Indonesian oil experts that both onshore and offshore areas of the northern part of West Java had a significant oil and natural gas potential (reserve). His pioneering and success as the first Production Sharing contractor between Pertamina and IIAPCO in 1966 in the Indonesian offshore areas of the northern part of West Java pushed the efforts for the exploration of oil and natural gas in other offshore areas, which in a relatively short period proved to be able to increase Indonesia's oil and natural gas production.

Ibnu, the most deserving of all, was the first to be presented this *Pelopor* commendation. His place in Indonesia's oil history was thereby publicly acknowledged and appreciated, even by his early detractors. I was very fortunate to have had the opportunity to be part of this exciting period of Indonesia's history. It was a wonderful experience.

APPENDIX

MAIN CHARACTERS

IIAPCO
    Pat McDonough—Stockholder
    Larry Barker—Stockholder
    Don Todd—Stockholder/Geologist/Negotiator
    Roger Machmud—Lawyer/First Salaried Employee
    Nancy Todd—First (and only) Non-Salaried Employee
CARVER-DODGE
    Dave Dodge—Partner/Geologist
    Doug Carver—Partner
    Bob Fowler—Lawyer
WARRIOR OIL COMPANY
    Joe Obering—Owner/Geologist
    Will Obering—Owner/Geologist
NATOMAS
    R.K. Davies—Chairman
    Chandler Ide—President
    Ben Tyran—Executive Vice President
    Keith Davis—Geologist
SINCLAIR OIL COMPANY
    E.L. Steiniger—Chairman & CEO

Fred Bush—President
Bob Elston—President Sinclair International
Bernie Moore—Chief Geologist
Ray Halsey—Chief Geophysicist
Ed Chittick—N.Y. Manager for Indonesia/Geologist
I.D. Stevens—Jakarta Manager/Geologist
Luther Horton—Geophysicist

INDONESIA
Sukarno—President for Life (Deposed)
Suharto—General/Successor President
Ibnu Sutowo—"Mr. Oil"/Permina
Anondo—Basic Mining/Permina
Wijarso—Basic Mining/Directorate of Oil & Gas Affairs
Pattiasina—Permina
Judo Sumbono—Permina
Sutan Assin—Permigan/Permina
Sanger—Lawyer/Permina Negotiating Team
Kwa (Tirto Utomo)—Lawyer/Permina Negotiating Team
Sakidjan—Economist/Permina Negotiating Team
Siregar—Accountant/Permina Negotiating Team
Bratanata—Minister of Mining
Soemantri—Successor Minister of Mining
Chaerul Saleh—Sukarno's Third Deputy Premier

OTHERS
Harold Hutton—Owner/Refining Associates
Tom Brook—Chairman/Asamera Oil
Andrew Alexeiev—Asamera/Singapore Representative
Howard P. Jones—U.S. Ambassador
Marshall Green—Jones's Successor
Frank Galbraith—Green's Successor
Paul McCusker—Counsel for Economic Affairs
Paul Cleveland—Commercial Attache
Bill Palmer—American Motion Picture Representative

The Philippines

Finding Oil Where It Shouldn't Be

By Allen G. Hatley

BALABAC CONTRACT AREA—GETTING STARTED IN THE PHILIPPINES

For about a year following mid-1974, Philippine-Cities Service Company's program in the Philippines had consisted of exploring the Balabac Service Contract Area in the Sulu Sea, southeast of Palawan Island. The Balabac area, although not very high on our new ventures list in the Singapore office for containing good drilling prospects, was an untested basin and the company had committed to drilling two test wells to earn its interest.

Our first commitment well was already due to be drilled. As a result, I had spent much of the last few months talking to various oil companies operating in Southeast Asia, attempting to acquire a 60-day "slot" on a suitable drilling rig. We planned to drill that first well, Coral #1, in almost 400 feet of water. About April 1975, I had finally arranged for, and our head office in Houston had agreed to, a "drilling slot" on the *Discoverer III* drill ship from Conoco. Conoco was currently using the "D III" rig to drill a well offshore Taiwan.

The Coral #1 well, with an expected total depth of 3048 meters (10,000 feet), was only the first of a two-well farm-in commitment made about a year previously in Houston. Cities Service was to pay 100% of two wells to earn a 50% Working Interest in Husky's Balabac Contract Area. Not only did we in Cities's Singapore and Manila offices believe the prospects had a very high geological risk, we also knew that the original farm-in negotiated in Houston with Husky Oil Company was not competitive. It was too tough a deal.

Superior had originally held the acreage, and several years earlier had shot seismic and dropped it without drilling because they did not consider it a prospect attractive enough to drill. After a thorough evaluation of the contract area, we in Singapore agreed with Superior's technical assessment of the area. That is not to say that the contract area was totally without merit; we believed, however, that more seismic and drilling than we could then justify would be necessary even to begin to determine where the best prospects might be located. As a result, if the first agreed-upon well in the basin turned out to be disappointing, we would know that we were in deep trouble because subsequent drilling "prospects" did not appear nearly as attractive as the first drilling location at Coral #1.

There was actually one good thing about the Balabac Contract Area: the area southeast of Palawan Island is a classically beautiful South Sea island setting. Otherwise, its hydrocarbon potential appeared rather gloomy. There were no predictable reservoir beds, no known source rocks, questionable thermal maturation, and even doubtful traps. It was an extremely high-risk play on which to commit to drilling two test wells.

Since the Balabac Contract Area had been acquired, we had managed our Philippine-Cities Service Co. exploration program from Singapore, where three years earlier I had opened a one-man new ventures exploration office. By mid-1975, it had grown to include five explorationists plus support staff. I also had a small operational office and staff in Manila, consisting of only one company expatriate employee, Jim Pate, an American accountant who was transferred there from Bogota, Colombia. The Filipino staff consisted of only six full-time employees, some of whom I had known for almost 20 years because I had been previously stationed in the Philippines from 1957 to 1964 with StandardVacuum Oil Company.

Because we planned to drill our first well in the Philippines within the next few months, our company's Indonesian office had loaned us a consultant, Jerry Roy. He was to prove to be the most competent and professional contract drilling superintendent I have ever met. During periods of drilling, we also supplemented our geological, drilling, and materials group with temporary assignments from other Cities Service operations.

Jim Pate, the "country manager," ran the office when I was not in Manila—which was about three out of every four weeks, except when we were drilling. Without Jim Pate and Jerry Roy, we could not have operated in the Philippines as successfully or as economically as we did. The local staff were among the very best: completely loyal and competent. We did not, however, have many good drilling prospects located in the Balabac Contract Area.

Another important part of the Philippines petroleum "infrastructure" is worth commenting on. That was the governmental group with whom we worked, and who were in charge of assuring our compliance with the terms of the service contract. The Petroleum Board, later to be renamed The Bureau of Energy Development (BED), was a small group primarily of technicians and attorneys, headed by Chairman Geronimo "Ronnie" Velasco and his assistant, Jose Leviste. Chairman Velasco was not the typical bureaucrat; he surrounded himself with predominantly competent and honest people, and he was pragmatic and accessible. Most important, he wanted to make things happen and knew how to do it. As a result, we always knew that if we had a problem, or a new idea on how to accomplish something, Velasco or his senior staff people (Jose Leviste, Art Sali, or Pat de la Paz) would listen.

In mid-1975, following our contracting of the *Discoverer III* drilling rig, the situation in the Philippines was actually the least of our company's problems. In Indonesia, the government's recent unilateral revision of existing Production Sharing Contract terms, and the probability that this would push other governments in Southeast Asia to "renegotiate" existing contract terms or stiffen new terms, was cause for real concern to our regional new ventures office. We were already actively evaluating several new exploration areas as far away from Southeast Asia as we could go, but still staying within our office's sphere of influence. This eventually turned out to be a good decision for the company, for we acquired interests in two new areas—the Badin area in southern Pakistan, and the Swan/Jabiru area on the Northwest Shelf of Australia—both of which would one day produce significant amounts of oil and gas.

The renegotiation of existing contracts by Indonesia's state oil company, Pertamina, eventually led to a slowdown in exploration in Indonesia and the other countries who opted for imposing less competitive contract terms on interested foreign oil companies. After 1979, rapidly rising oil prices caused many groups to accept the more onerous contract terms then asked in Asia and elsewhere. But for a regional new ventures manager in Singapore in 1975, this appeared to be the most pressing problem our company faced at the time.

THE BIG SURPRISE: A DRILL SHIP NAMED THE TAINERON

A very different problem surfaced one afternoon in May 1975, however, when I received a call from Les Wood, who identified himself as a representative of Texas-Pacific Oil Company, the prime contractor for something called the Drill Ship *Taineron*. Les asked if I could see him the next day to discuss "where we [were] going to use the *Taineron*." I was convinced the caller was mistaken, for Cities Service's Houston management had never mentioned the *Taineron* to me. Furthermore, I had just spent several months searching for a drilling rig and only recently signed a contract to drill the Coral #1 well in the Philippines.

At the meeting the next day, Les Wood described the D.S. *Taineron*, which was then being built in Hong Kong, along with the terms of a drilling contract agreed to by Cities Service in Houston some three months earlier.

Les Wood advised that Cities Service had committed itself to take three drilling "slots" of at least 60 days each, at a day rate estimated at $40,000/day. This meant that, using an estimated 60 days to drill and test a well, the cost per well would be about $2,400,000 in contract drilling costs alone. Adding another $2,000,000 in equipment and supervision costs, we appeared to be committed to spending an estimated $13,000,000 in drilling three wells somewhere in East Asia, over the next 18 months, using the D.S. *Taineron*. This was among the most expensive drilling rig day rates in East Asia at that time.

The first problem was that we had no viable prospect to drill in the Philippines, unless Coral #1, which would begin drilling in about 60 days, was successful. Drilling a wildcat well in an untested basin and expecting that first well to be your only justification for quick additional drilling is a risky enterprise at best. We also had no other acreage approved for acquisition anywhere else in Southeast Asia, even though the D.S. *Taineron* was scheduled to be commissioned in just five months.

We quickly confirmed that the drilling contract actually did exist. A few weeks later we visited Hong Kong to see the D.S. *Taineron*, to get an idea what it looked like, and to determine what we could expect of the rig. In June 1975, I returned to Houston to discuss our forthcoming drilling program at Coral #1. I also discussed several new areas in which the new drill ship might be used. During my short visit, it was my understanding that because of the manpower they had available, our Houston office would lead the search for new areas in which to use the D.S. *Taineron* rig.

We spudded the Coral #1 well on August 13, 1975, and drilled a monotonous shale and siltstone stratigraphic section to 3062 meters (10,047 feet). No hydrocarbon shows were encountered, and it was almost warmer on the rig floor than it was at total depth (155°). It was a cold, dry, and disappointing wildcat well, but that was not really unexpected.

During the drilling of the Coral #1 well, the Cities Service Singapore new ventures office continued to look for new areas in which to use the D.S. *Taineron*. We also tried to engage Houston in a meaningful dialog, but it was as if no one seemed to be listening. By late September 1975, and following the plugging of the Coral #1 well, we were becoming somewhat desperate, and the only suggestion we had received from Houston was that maybe the company could use the D.S. *Taineron* in Cities Service's Java Sea contract area in Indonesia.

What had been forgotten by those proposing the ship's use in Indonesia, fortunately or unfortunately depending on how one views future events, was that Pertamina had placed a cap on the dollar cost of drilling rigs that could be used in Indonesia. This was because of the government's desire to hold the "cost oil" accounts to a minimum. The D.S. *Taineron* day rates far exceeded that cap.

Once we had completed the Coral #1 drilling program in the Balabac Contract Area, both Cities Service's Manila office and its Singapore office went back to the drawing board and agonized over where to use the D.S. *Taineron*, either in the Philippines or elsewhere. A second well in the Balabac Contract Area was never an acceptable choice, because of the lack of a good geological drill site. Where else we should go, or could go, had not been answered yet.

ON OUR OWN: A GREAT OPPORTUNITY—MAYBE?

During October 1975, I returned again to Houston to discuss a number of new venture projects we were working on in Singapore, and to discuss the possibility of further drilling prospects in the Philippines. The future use of the D.S. *Taineron* rig was high on the list of projects we discussed. Houston seemed still to have no new ideas where the rig might be used within the overall company organization, nor were they outwardly very concerned.

While in Houston, I finally accepted the obvious, which was that our Singapore office needed to assume complete responsibility over where this rig would be used, and, therefore, where approximately $13 million would be spent in exploration in 1976 and 1977. During the Houston visit, it also became obvious that we in the Singapore and Manila offices were completely on our own and had to move ahead aggressively in order to find a better prospect than a second well in the Balabac Contract Area, which was our only current drilling possibility.

For an explorationist, this was a great opportunity; but for the manager of small offices in Singapore and Manila—suffering from continued benign neglect from the home office—it would, over the next few months, be very disturbing. By now the D.S. *Taineron* was on "sea trials" and we estimated we might be only about 100 days from its delivery.

Due to the problems inherent in completion of negotiations for any new contract area in the region, we just about had to limit our search for a new exploration play, and the first drilling "slot" on the Drill Ship *Taineron*, to the Philippines. This was because we thought we could make an agreement quicker in the Philippines than in any other country of Southeast Asia.

At about that time, I recalled that sometime back in 1973, six months or so before the Balabac Contract Area had been acquired, I had been in Manila discussing possible acquisition opportunities with several local oil companies. During the course of these discussions and data reviews, I had been shown several seismic lines covering an area offshore northwest Palawan Island, held by a local group, Oriental Petroleum & Minerals Corporation. In 1971, this small company had actually drilled the first two wells (Pagasa #1 and Calamian #1) offshore the Philippines. The seismic lines that I had looked at covered an area around one of these wells.

Both of these 1971 tests were dry holes, containing little or no encouragement, except *possibly* mature source rocks near total depth. No reservoir rocks or hydrocarbon shows had been noted and, more importantly, I found that less than 30 meters (100 feet) of dense carbonate rocks had been found in the wells. This was important because I had thought at the time that I reviewed the seismic work that I had seen indications of possible reefal buildups in this area. If there were reefal carbonates on the seismic lines, this meant that neither of the previous wildcat wells had actually tested what might be the very attractive play in this offshore area.

By the time I was ready to leave Houston, I had decided to reexamine this new play, which I felt might be an area of untested Tertiary age carbonate reefs located in an ideal geological setting offshore northwest Palawan Island. If the play was actually there, we should attempt to acquire these rights. There was still another major problem, however, since already a number of other wells had been drilled elsewhere west of Palawan Island. None of these wells, so far as I knew, had drilled a carbonate reef facies—not even the two wells drilled in 1971 in this same vicinity. Had I been dreaming, or was there a untested carbonate reef play in the area?

I had long been convinced that carbonate rocks were the reservoir of preference in the Philippines. In fact, in 1971, a friend, Ed Durkee, and I had co-authored an article for the *Oil & Gas Journal* wherein we recommended carbonates, and reefs in particular, as being the main reservoir objectives in the Philippines. Nothing I had seen since 1971 convinced me that we were wrong in that evaluation. Yet apparently no foreign oil company had reviewed this seismic data, and if they had, they had not recognized the importance of reefal buildups in this area offshore northwest Palawan Island. In my opinion, if the play was there, that was the area we should acquire and drill. Before leaving Houston and returning to Singapore, I contacted Jim Pate in Manila to determine whether several local companies still held clear title to the offshore leases around northwest Palawan Island.

The answer to that question was in Singapore awaiting me upon my return near the end of October, 1975: the leases were still available, but other groups "were said to be talking" to Oriental Petroleum & Minerals Co., the main Filipino concessionaire, about these leases. I called Oriental and got their approval for a further review of their technical data in Manila the following Monday. Les Beddoes, probably the best explorationist I have ever worked with, and my Exploration Manager in Singapore, and Herb Drushell, a fine geophysicist, flew to Manila for the review. [It is worth mentioning that throughout only a 5-year period of working with Les Beddoes, we acquired interests in three different countries, including the Philippines, in which more than 450 million barrels of oil have been discovered. Beddoes is one of the best.]

Les Beddoes called me from Manila two days later and confirmed that he and Herb Drushell agreed that there were several reefal buildups located at the top of a thick carbonate section appearing on a number of the seismic lines they had reviewed. They were prepared to recommend an acquisition. I flew to Manila on November 4, 1975, and obtained an "option" on the concession rights of several Filipino-owned oil companies who held leases in part of our new area of interest. There were a total of four locally owned groups already holding interests: Oriental Petroleum & Minerals Corp., Basic Petroleum & Minerals, Inc., Philippine Overseas Drilling & Oil Development Corp., and Landoil Resources Corp. In addition, we planned to include a large open area in the South China Sea contract application.

There was, as I mentioned earlier, an additional consideration and that was that I had not talked with Husky Oil Co., who were the original lessors in the Balabac Contract Area, about a possible "transfer" of our drilling obligations for a second well to this new area. This was because their local manager, Jim Manning, was out of the country. While it was important to eventually contact Husky, I was confident that Husky was also very disappointed in the results at Balabac. As a result, I felt Jim Manning would agree to move our obligations to a new and geologically more attractive area in northwest Palawan. All of these plans would be, of course, no use to us in moving the second well obligation from the Balabac Contract Area if the government of the Philippines would not also approve such a scheme. I personally felt there was a good chance they would agree to it and while in Manila, I also approached the Petroleum Board on the basis of a hypothetical scenario. They would not agree, but did not turn us down at the time.

That was all we needed. We decided to move forward and recommend the area for acquisition. At worst, the northwest Palawan area might offer us a good drillable prospect for the second or third "slot" on the D.S. *Taineron*, if we could not put it together in time to replace a second test well at Balabac. The D.S. *Taineron* was by now completely built, had just arrived back in Hong Kong from "sea trials," and was about to go to Texas-Pacific Company to drill at least one and we hoped, two wells off the east coast of the Philippines. We now estimated we had only about 60 days before the rig sailed into our view.

On November 4, 1975, it was time to contact Cities Service in Houston with the good news that (a) we had an exciting prospect to drill in the Philippines, and (b) the D.S. *Taineron* would be put to good use. We sent Houston a 3-foot-long telex with our recommendations, expecting to have their quick approval, after a few exchanges of questions, so we could get on with the acquisition. We were greatly mistaken. Fifteen days later, and with the D.S. *Taineron* then drilling its first well for Texas-Pacific off the east coast of the Philippines, we were still waiting for our Houston office to approve our recommended acquisition of the northwest Palawan area.

SOME BACKGROUND—IT HELPS TO KNOW THE PLAYERS

The Philippines, along with Canada, Australia, and the United States, are among those unique nations who have allowed a private oil sector to flourish. Most other nations in the world have controlled the exploration for oil and gas within their boundaries by instituting strong governmental constraints on exploration rights. In the Philippines, since about 1938, small, locally capitalized oil exploration companies have been active on the Manila Stock Exchange. They have also been active in acquiring leases and in drilling wells, all but two of them onshore, and mostly drilled shallow. However, in 1975, after more than 100 wells had been drilled for oil, no commercial oil discoveries had been made in the Philippines.

Up until the mid-1960s, a number of the world's large oil companies had explored in the Philippines, including: Caltex, Standard-Vacuum Oil Co., Exxon, Mobil, Chinese Petroleum Corp., and Union Oil of California. With the exception of a good, but remote, dry gas discovery by Standard-Vacuum Oil Company in 1958 (I had been a well-site geologist on this well), nothing really had come close to being a commercial hydrocarbon discovery.

The history of petroleum exploration in the Philippines has also had its share of "booms and busts." Just after the end of World War II, a few wells were drilled on the island of Cebu. Oil exploration activity was, however, most intense from about 1956 to 1964. Most of the local exploration companies went belly-up after that and their stock became almost worthless. About 1970, as offshore drilling took hold in other nearby areas of Southeast Asia, a "mini-boom" in new offshore leasing occurred in the Philippines. Numerous local oil exploration companies were formed by those same people who had formed the previous, now defunct, companies about 15 years earlier. Never mind their previous lack of success, somehow they sold their dreams of discoveries and shares of stock from the same place, the Manila Stock Exchange.

But the years from about 1964 to 1970 had not been kind to the Philippines. Riots, an indifferent government bureaucracy, a largely corrupt Congress, an irresponsible press, a more active communist insurgency, a revolutionary "Holy War" in the Muslim areas in the southern Philippines, and a seriously floundering economy had greeted me upon my own return to Manila in 1969, after having left that beautiful country about 5 years previously. The changes during those 5 years were remarkable and obvious. I remember to this day someone asking me in 1970, "What did you think of the Philippines when you went back?" I replied that I had "looked into the eyes of a lot of people who had seen better times—and they knew it !" I had that kind of an impression.

There were, therefore, the makings for all sorts of revolutions and chaos. Greed and corruption were alive and well in the Philippines in 1970. A few years later, in 1972, "Martial law" was declared by the then-elected president of the Philippines, Ferdinand Marcos, in an effort to solve some of these problems. Some 15 years later, in 1986, most would criticize this action, but martial law was effective and probably the least of all the evils available to those trying in the early 1970s to either hold onto or to change the old system of government and patronage in the Philippines.

Outside influences were also affecting the Philippines in the early 1970s, including the 1972 Arab oil embargoes. While oil price increases hurt us all, if it had not been for the Arab embargoes of oil in 1972, the Philippine exploration "mini-boom," which had sputtered along for a couple of years, would have resulted in no more than an additional number of bankrupt local oil companies. Real activity—drilling—usually happened in the Philippines only after a local company with leases teamed up with a foreign group who had money and who wanted those leases. But this time there was a difference. After decades of foreigners exploring for oil, copper, coal, and other minerals, and acquiring property, the Supreme Court of the Philippines in the 1960s had "reinterpreted" its 1935 Constitution and declared it illegal for foreigners to own a majority interest in land or minerals, including oil leases.

As a result, no foreign oil companies had shown up at the "mini-boom" to bail out the Filipino companies. Oriental Petroleum & Minerals had been forced to scratch up its own drilling money in 1971, in a series of "private placements." They were able to get two "cheap" offshore wells drilled, but no oil was discovered. We later suspected that maybe the wells were not even drilled exactly where they were mapped. The only thing that had been acquired was some good seismic data, and this we had just finished evaluating.

The Arab oil embargoes also helped to convince President Marcos that a country like the Philippines, who depended on importing about 85% of its energy needs, could ill afford to cut itself off from foreign investors in the oil exploration game. Marcos, who had recently declared martial law in the Philippines, used these powers in late 1972 to reform part of the way oil exploration was undertaken in his nation. A "Service Contract," essentially a modified Indonesian Production Sharing Agreement, was declared the "Law of the Land," and foreign companies were again allowed to conduct exploration under this type of contract. The hunt for Philippine oil was finally on!

As a result, between late 1972 and 1975, Chevron, Texaco, Sun, Phillips, Arco, Champlin, Texas-Pacific, Superior, Husky Oil, Cities Service, and Amoco, either alone or with numerous local oil companies, acquired leases, and some even drilled test wells. But still, no oil was discovered. Although this was an impressive list of large multinational companies, there were still very few aggressive, large oil companies exploring in the Philippines. This was primarily due to the fact that by 1975, there had not been an oil or gas discovery in all of the Philippines, from Taiwan in the north to Borneo, nearly 1200 miles to the south. If you asked a lot of good geologists at the time, most would tell you, "The Philippines lacks good oil source rocks, reservoirs, and effective traps. There are better places to look for oil and gas!" As a result, although there were many big-name oil companies represented in the Philippines, most were there temporarily, just "following the crowd."

ON WITH THE HUNT FOR OIL

After we recommended to Houston on November 4, 1975 that Cities acquire a Service Contract in northwest Palawan, and before the recommendation's approval, we began to tie up some loose ends. A meeting with the Petroleum Board's executive director, Jose Leviste, was very satisfying. We were advised that the government (meaning

Chairman Ronnie Velasco) agreed, in principle, with our idea of "transferring" our second obligation well to a new block in the northwest Palawan area. We also began to discuss Service Contract terms and to determine work and expenditure commitments, although still only on an unofficial basis, while we awaited approval of this acquisition from Houston.

Jim Pate and I also met with the heads of the Filipino companies and worked on both the agreements between our companies and on the Service Contract itself. During this time we received our first reaction from Houston to our recommended acquisition and description of the geological play. It read as follows: "WE CONTINUE TO BE OVERWHELMED AND JUST WAITING FOR THE OTHER SHOE TO DROP," followed by questions about what we would earn in the Balabac Contract Area, if we did not drill another well. Not very satisfying, but we rationalized that it must have been a lot of new ideas for our Houston office to contemplate, especially since they never had seemed to realize in the first place that there was very little of value to earn at Balabac.

For the next 15 days we traded some telexes and telephone calls with Houston. Economics of the new Service Contract, based on verbally agreed-to terms, looked very good. Husky had finally agreed to all we suggested. We were only awaiting approval by Cities Service's main office so that we could sign a final agreement with the Filipino companies and officially agree to the draft Service Contract. But still no approval came, and the D.S. *Taineron* was, we believed, nearing total depth on the Texas-Pacific well.

In the middle of November, it became clear to both our Singapore and Manila offices, after having received no advice or agreement to our recommendation for over two weeks, that "normal Cities Service channels" were just not working.

A few days later, on November 20, 1975, at about midnight Manila time, I finally called P. W. J. (Jim) Wood, Vice President for International Exploration and Production for Cities Service in Houston; it was late morning in Houston. Jim Wood was not my direct supervisor, but he was capable of understanding that there was a problem with the remaining prospects in the Balabac block and that we needed a new and more attractive area to explore. Jim and I had also previously worked together in two other companies, and he had originally hired me to open the Singapore office for Cities Service. Most important, however, was the fact that he was also the only person who would or could approve the prospect and acquisition ideas without weeks of further delay. We still faced the probability that the *Taineron* drill ship would arrive before we could put the deal together and get our location surveyed.

Jim and I talked about everything, including the geological plays, the terms of the new contract and its economics, our commitments with Husky, the Philippine government's attitude, and the *Taineron* rig. Jim Wood, of all the individuals present when I had been in Houston the month before, had even seemed to understand the problems. In addition, he was a geologist and he accepted the concept of "exploration risk," which is often not well appreciated. It was, in my opinion, a concept always in short supply in Cities Service. Within about 40 minutes that night, he and I reached a general agreement and he agreed that, as soon as practical, he would telex approval to go ahead with acquiring the Northwest Palawan Service Contract Area.

During the conversation with Jim Wood, I specifically asked for the help of Rod Buckles, the head of the International Legal Department, and Leonard Milligan, a top-notch petroleum negotiator. They were to come to Manila as soon as possible in order to take over the details of the negotiations and finalize an agreement with the Filipino companies. At the same time, we also had to obtain final government agreement on a service contract and prepare for the arrival of the D.S. *Taineron*.

Within 24 hours, we received telex approval to acquire the Northwest Palawan Service Contract Area. Milligan and Buckles arrived just four days later to complete negotiations and finalize legal documents. Things looked pretty good for a change, but there were hundreds of things to do and problems to solve before we had the new contract area and a new drill ship safely located on the right drill site.

In Singapore, Les Beddoes and his exploration group began to integrate the new geological and geophysical data we had acquired in Manila with the other information we already had from the same area. This would lead us to designate an interpreted Oligocene-Miocene reefal buildup as our first drill site, Nido #1. It was also one of the same reefal buildups I had seen in Manila almost 18 months earlier, while reviewing seismic and well data.

Somehow this play had sat in Oriental Petroleum's files for more than 4 years. The area had been reviewed by a number of other groups, none of whom had acquired it. While this may be hard for an outsider to understand, that's not uncommon when groups just "follow the crowd" from one country or basin to another, and fail to originate and fully understand geological play-types or even dig for data to review.

In Manila, Leonard Milligan and Rod Buckles completed negotiating the agreement with the Filipino partners. I signed it on December 2, 1975, and we immediately sat down with the government to finalize the new service contract. We had already verbally agreed to most of the variable terms of the service contract, and on December 5th, the Petroleum Board's Director Velasco approved the contract and sent it to the president's office for approval.

A few days later, we found out that Texas-Pacific would need several more weeks to complete drilling its well, before they would release the *Taineron* drill ship to Philippine-Cities Service. That meant we still had at least 30 or 40 days before the rig would be on our new contract area. We were lucky. Our Manila office's priorities quickly turned toward organizing and supplying those items that we would need for drilling a 2743-meter (9000-foot) offshore wildcat well in 113 meters (372 feet) of water.

Fortunately, there was little change in requirements, except bits and some mud materials, from those supplies that had been loaded on the D.S. *Taineron* in October before it sailed from Hong Kong. Our timing for the drilling was also accidentally perfect, since the typhoon season was just finishing. Our most immediate need was for a sea bottom survey at the Nido #1 location.

On December 17, 1975, Philippine-Cities Company, as operator, along with the four Filipino companies, one Canadian company, the Philippine National Oil Company, and the Philippine government, signed Service Contract No. 14, consisting of 515,600 hectares (1,274,047 acres) offshore northwest Palawan. It was hard to believe that it had been only 26 days since we had obtained head office approval and had approached the government of the Philippines with the idea of acquiring a new service contract area and transferring our second well commitment to that contract. In retrospect, it was truly a remarkable accomplishment that could only have happened after martial law and the organization of the Philippine government's Petroleum Board, and its successor, the Bureau of Energy Development.

DRILLING AND TESTING THE NIDO REEFS

On January 31, 1976, we spudded Nido #1, in 113 meters (372 feet) of water, using the Drill Ship *Taineron*. Wes Hicks, a geologist formerly stationed in Indonesia and Singapore, was now assigned to the Philippines operation, along with Matt Reber, on loan from Cities Service Indonesia, as Operations Manager. Otherwise, we again used all consultants and contract labor. We had, as the saying goes, "come a long way," but we were still not in the "main stream" with Cities Service.

Drilling ahead at 1817 meters (5960 feet) we encountered a porous carbonate reef with oil shows. Contrary to what we did in later wells, no coring or drill-stem testing was accomplished in Nido #1, in large

Left to right: Jim Manning, Husky Oil Manager; Allen Hatley; and Jim Pate, Philippine-Cities Service Country & Finance Manager, in Manila, 1976.

part, because of hole conditions. At 1878 meters (6160 feet) we lost complete returns. From there to total depth we had only partial circulation. With our fingers crossed and, I suspect, lowering the South China Sea by several feet as we drilled ahead using mostly sea water, we drilled to 2751 meters (9026 feet), our total depth.

After plugging back to 1826 meters (5991 feet) and acidizing with 3000 gallons of 15% HCl, we tested and flowed an estimated 1400 barrels of 27° API gravity oil per day, on a 24/64" choke, from 5 meters (17 feet) of net porosity. It was an exciting day, with the roar of that much oil being produced and burned to avoid pollution.

The day following the testing of Nido #1, I took the long helicopter ride back to Puerto Princessa on Palawan Island and then flew by Philippine Air Lines to Manila. Upon arriving at the hotel, I turned on the TV and unpacked. As the TV set came on, the first thing I heard was "Nido," and looking at the TV screen I saw an aerial view of the *Taineron* drill ship. A long string of black smoke floated away from the ship as oil had been burned during the testing the day before.

The telephone rang, and it was the office of the Chairman of the Petroleum Board, Ronnie Velasco. They said, "Chairman Velasco is on his way to pick you up. You are going to meet President Marcos." I jumped into and out of the shower, dressed, and as I went downstairs, I pulled from my briefcase two bottles containing some of the first oil produced at Nido #1. Velasco was waiting at the front door when I walked out, and away we went to Malacanang.

Malacanang Palace originally had been the home of the Spanish and then the United States Governor General. Since 1936, it had been the official residence of the President of the Philippines. We waited our turn to meet the president, along with about 20 other people, in groups of three or four. All of us were there to personally report on this or that project, or to seek funding and patronage for some idea.

Finally we were ushered into the office of the president, who was first briefed by Velasco and then by myself as to what had happened at Nido #1 during the testing. After about 15 minutes, Mrs. Marcos entered the room and congratulated Cities Service on its success. Mrs. Marcos had, at about that same time, been made Mayor of Metro-Manila. This thankless job consisted of trying to coordinate the efforts and control the egos of the elected mayors in the dozen municipalities surrounding Manila.

During the hour or so I spent with President Marcos, I suddenly realized how much of an impact Nido and our future efforts would have on the Philippines, and should have on Cities Service. This meeting with President Marcos was only the first of some eight or ten meet-

The Drill Ship *Taineron* conducting the first production test at Nido #1 in the Philippines.

ings we would have over the next 14 months. All proved valuable to the process of getting "our story" to the right man, and giving us a better idea of the problems faced by the government.

On March 13, 1976, Nido #1 was plugged and abandoned, as are most offshore wildcat wells. The Drill Ship *Taineron* sailed away to the Gulf of Thailand to drill another wildcat well for Texas-Pacific, and we began to plan our next test well.

It is worth mentioning that the government of the Philippines, the Manila Stock Exchange, and the Southeast Asian oil community had never seen anything quite like Nido, and neither had we. Philippine-Cities Service Inc. was initially the "idol" of Manila, but we were not actually trusted by everyone. After so many companies had tried to find oil and failed, it was hard for many to believe that we had succeeded. As a result, remarks from more than one commentator at the time went something like this: "Surely Cities Service had the government report the oil flow just to make the local oil company stocks go up." Although by and large the Filipino oil company entrepreneurs and corporate officers are really very good people, this association with Filipino stock companies would often in the future significantly impact our program, and to some extent it would affect how we did business in the Philippines.

During the drilling of Nido #1, we recommended that a new seismic survey be run over the Northwest Palawan Service Contract Area. That survey began on March 27, 1976. A few weeks before this survey began, and before we had any new information on the other prospects in the Nido area, stories began to appear in the Philippine press about the development and production of the discovery at Nido #1. At first I thought this was just so much local stock promotion.

In a meeting with the government representatives, we had explained that Nido #1 appeared to be too small to be produced alone and that it was our intent to await the results of the seismic survey to determine the next (and, hopefully, better) prospect to be drilled. I was told that while the Petroleum Board did not disagree with our assessment, they thought it was only a matter of time before our Philippine partners proposed the development of Nido #1 with government assistance.

Their reason was, of course, the Manila stock market. Some in our group suggested we let them go ahead and spend their limited funds on developing Nido #1. By doing this they would then have to drop out of future and potentially more profitable wells. On the other hand, I always felt that our group ultimately would be stronger if it had good Filipino partners, and I began to search for a way to separate them from Philippine government support.

It soon became clear that several of the stronger politicos in the Filipino partners were lobbying with President Marcos for development of a very marginal discovery at Nido #1. I asked Ronnie Velasco to arrange a meeting with Marcos in order to try to head off a "no win" situation for all concerned. Within a few days, Velasco advised me that "The Boss" would see me, and I arrived at Malacanang with maps, drilling logs, and a one-page summary of test results from Nido #1. President Marcos, Velasco, and a few others listened to my analysis that Nido #1 was a discovery too small to develop alone. Furthermore, I explained that we were just about to shoot a seismic survey to learn what the potential really might be for northwest Palawan. I also stated that what I was saying was probably disappointing to the president, but a decision to develop Nido #1 at this time was a bad decision for all concerned.

At the end of the discussions, I said that I had no official authority to make the offer, but that I thought if the Filipino partners wanted to go ahead with developing Nido #1 at this time, Cities Service would probably prefer to carve out a one-square-mile block around the discovery, assign it to the partners, and Cities would simply walk away from it. As Velasco and I were leaving, the president asked if I intended to be in Manila over the weekend. I replied that I did. Marcos then said, "Allen, meet me at Pier 3 on Saturday morning at 8:00 AM, and we will go out and visit Corregidor with some friends."

On Saturday, not knowing exactly what to expect, I arrived at Pier 3

Left to right: Allen Hatley and Petroleum Board Chairman Geronimo Velasco present President Ferdinand Marcos with a sample of oil from the Nido #1 well.

Left to right: Arthur Saldivar-Sali, Deputy Director of the Petroleum Board, Jim Pate, Geronimo Velasco, and Allen Hatley meet with President Marcos to discuss the partners' desire to develop the Nido #1 well.

and found about 30 people, along with President and Mrs. Marcos, boarding a large ocean-going yacht. Twice during the trip, I had an opportunity to talk with the president about the oil business in Southeast Asia. We returned to Manila late in the afternoon, after visiting the Bataan Peninsula and Corregidor. I left Manila two days later for Singapore.

Early the next week I received a telephone call from Chairman Velasco, who told me that "The Boss agrees with you, now is not the time to waste our money" on Nido #1. It was a small victory, I thought, and a decision Marcos and the Philippines could ultimately be proud of making.

Our seismic survey was completed by April 10, 1976. For the next several months we processed and interpreted those data. The up side was that we had several prospects to drill and, based on the results of Nido #1, we were very confident we would find a larger oil accumulation. At the same time, we had one basic "oil-finders'" quandary to answer before we could successfully exploit the northwest Palawan area.

We had learned that Nido #1 was drilled on a rather small reefal

Finding Oil Where It Shouldn't Be 137

Testing South Nido #1, the discovery well at the Nido Field, 1977.

buildup located on a large carbonate shelf. We had similar but somewhat larger "reefs" up dip and down dip from the Nido #1 reef test, the analysis of which had told us was only partially "full" of oil. We also had a few "monster" basement structures, located mostly up dip, possibly with porous carbonate shelf deposits draped across them. We felt they were all good prospects, but which was the "best" location for a second drilling?

We finally decided to follow the advice that all geologists are told in school and that we usually find corroborated in our work, which is to drill the next well "up dip from a good oil show." It proved the wrong thing to do. The North Nido #1 test well, drilled in late 1976, was a dry hole, with only a few oil shows and little porosity or permeability.

A second, and larger, seismic survey followed. This seismic survey better defined the prospects available in the contract area. For the first time, we had seismic data from the northern part of the contract area, where we saw even more "reefal" prospects and got a better idea of the regional geology.

If we couldn't find large oil accumulations up dip from a good oil show, then maybe, we reasoned, we should drill down dip. Furthermore, we felt that maybe the reason that Nido #1 had only 5 meters (17 feet) of net oil column was because our primary source of oil was down dip, and filled our reefoid buildups to the spill point, before migrating further up dip. Thus we elected to drill South Nido #1 as

Nido Field development—production and well platform installation in the foreground, and a second well platform in the distance, with wells being drilled.

the third wildcat well in the Northwest Palawan Service Contract Area.

On June 6, 1977, we spudded South Nido #1 and drilled to a total depth of 2441 meters (8010 feet). After acidizing with 2600 gallons of 21% HCl, the carbonate buildup flowed an estimated 10,284 barrels of oil per day from 75 meters (246 feet) of oil column.

It was impressive! It was also the beginning of the discovery and development of a half-dozen small, but prolific, "Nido-like" reef discoveries (South Nido, Matinloc, Cadlao, Pandan, Libro, and Tara) in the Philippines. All except Cadlao were discovered by the Cities Service consortium. Initially it was estimated that this geological play contained about 175,000,000 barrels of oil in place, and had an estimated primary recovery of only about 20% of the oil in place.

To date, almost 35,000,000 barrels have been produced from the oil fields in the Philippines. By international standards, the Nido reefal carbonate discoveries offshore northwest Palawan represent comparatively small, individual oil discoveries. But for all of those participating in the discovery, it was significant and very profitable. Exploration and development costs were recovered in less than 12 months.

For the Philippines, Nido was the first commercial oil discovery

Captive tanker at Nido Field in the Philippines, acting as a floating oil storage facility.

ever made in the country. It also came at a time when oil prices were about to double, and the first discovery was primarily responsible for spawning continuing new attempts to make the Philippines self-sufficient in energy.

For the operator, Cities Service Company, Nido represented the first commercial oil discoveries the company made outside of the Western Hemisphere, after almost two decades of international exploration. For the half-dozen or so small Philippine oil exploration companies who participated in the program with Cities Service, it represented an opportunity to use the cash flow from Nido both to reward their shareholders and to gain some independence in future oil exploration efforts.

For the oil finder, it was the sort of experience you always want to duplicate, yet you know that no new find will ever take the place of the Nido discovery. And you are also absolutely sure that only a very few observers can really understand how it feels.

EPILOG

More than 15 years have passed since the first Nido discovery. Cities Service and Husky are gone. In late 1985, Occidental Petroleum, who had acquired Cities Service interests in a 1983 merger, sold Nido and the other oil fields, along with the undeveloped exploration interests it had acquired from Cities Service, to a small but aggressive group called Alcorn International. Chevron also sold the Cadlao field to this same group. Enough time has passed that even the control and direction of most of the small Filipino oil companies in the Northwest Palawan Service Contract have changed.

Other than another half-dozen similar reefal buildups having been discovered in the Philippines, the most exciting exploration development occurred about 10 years ago when Cities Service drilled an unexpected porous turbidite sandstone reservoir at Galoc #1, in over 305 meters (1000 feet) of water. Galoc tested 1712 BOPD in the same contract area as Nido. Occidental Petroleum also sold the Galoc well to Alcorn, and although it was never developed, more than 400,000 barrels of oil were produced by Alcorn during an extended testing period in 1989. Proven reserves have nearly tripled in northwest Palawan, as Alcorn, Philodrill, and Petrocorp have drilled successful new test wells.

Occidental Petroleum "rediscovered" northwest Palawan in 1989 when, in a new work contract area covering part of the Cities Service acreage that they had dropped four years earlier, they drilled a large natural gas discovery in over 610 meters (2000 feet) of water. This gas discovery was originally estimated to contain perhaps 2 or 3 tcf of gas. A second well drilled nearby was not productive. At that water depth, and with no infrastructure in the Philippines for the use of natural gas, time will tell whether additional reserves and a market can be proven.

Are the Nido discoveries unique? I believe they are, because more than 15 years later, there still are no discoveries of commercial oil or gas fields outside of northwest Palawan in the Philippines. *But let no one fool you—there is more oil to be discovered in the Philippines.* I suggest that the next time an oil finder is sufficiently challenged, and finds a way to get his wells drilled in the Philippines, there will be another discovery.

Exploring Onshore England

The Story of Marinex and the Humbly Grove Discovery

By John C. Kinard

INTRODUCTION

The Humbly Grove oil field was discovered in southern England in 1980, at the height of the "oil boom." This discovery was significant in the U.K. onshore, as it followed years of low activity that had resulted in only one major find. Although the early predictions of 15 to 20 million barrels of recoverable oil were never realized because of the poor quality in much of the Jurassic reservoir, this discovery acted as a catalyst for more significant oil discoveries in the future.

The background leading up to this discovery involves many interesting events and stories that are only a small chapter of the continuing quest of explorationists to find oil and gas reserves throughout the world. I was fortunate to play a small part in the revival of southern England exploration and the resulting industry successes.

The story begins in 1971, when a small private English company, Marinex Petroleum Ltd., was formed by three ex-Phillips geologists who had worked together in Montana 15 years earlier. This new company would ultimately be the vehicle that served as the nucleus for the consortium that discovered Humbly Grove as well as other oil fields in southern England.

SOUTHERN ENGLAND'S EXPLORATION HISTORY

The U.K. oil industry dates back to 1694, when a patent was taken for "oyle," a liquid obtained by separating tar from shale by distillation. However, the first significant activity was the "Scottish oil shale boom," which began in the 1840s and peaked in 1913 with the cumulative production of about 1.5 million barrels of oil. This oil was extracted from coal and oil shale and was used primarily for lubricating, wax, and kerosene lamps.

Of the more than 400 wells that have been drilled in the U.K. onshore, only about 100 have been in southern England. The first encouraging well was drilled in 1895 as a proposed water well near the Heathfield Hotel in East Essex County. It encountered and produced gas in the shallow Wealden formation (Cretaceous). Not until 1959, however, was oil discovered—in the Jurassic at the Kimmeridge field on the Dorset coast southwest of Bournemouth. This discovery, although small in areal extent, has produced more than one million barrels of crude oil and was a clue to the future. The well site is impressive—it could be viewed as a lone pump jack (or nodding donkey, as they are described in England) on a cliff overlooking the English Channel. In 1964, a small Jurassic discovery was made 10 kilometers (6 miles) to the north but was later shut-in. Esso drilled a BP (British Petroleum) farm out in 1965 that encountered a small Jurassic (Corralian) gas field at Bletchingley. During the 1960s and 1970s, only one to five onshore exploration wells were drilled each year in southern England. These were all dry holes except those previously mentioned and the more significant Wytch Farm field, which was drilled in 1974 but was held highly confidential. This graveyard of dry holes and low reserve finds set the stage for the activity generated by the Humbly Grove discovery in 1980.

During the lull of onshore drilling in the 1960s and 1970s, attention was drawn to the North Sea basin, which was being revealed as a major source of oil and gas reserves. In the Netherlands, the Groningen field was discovered in 1959 and deepened in 1963 to uncover giant gas reserves (86 tcf). Oil was first discovered offshore in the North Sea in 1969 at Montrose, and within a year the giant discoveries at Ekofisk and Forties (1.5–2 billion barrels of oil) followed.

THE MARINEX FOUNDERS AND FORMATIVE YEARS

My earlier career had been as an exploration geologist for Phillips Petroleum Company in Billings, Montana. While at Phillips I had worked with two other geologists, Dan Williams and Andy Fish, with whom I would later join in forming Marinex. Billings was an enjoyable small city in the 1950s, with a profusion of young geological talent that was part of the postwar effort to find more oil. Later, as the oil companies began one of their many consolidations or down cycles, much of this talent scattered. I resigned in 1960 to become a consultant and independent producer in Billings. Dan Williams was transferred to the international department of Phillips, and Andy Fish was transferred to various U.S. offices of Phillips. In 1970, Andy and I founded Compex as a private Nevada Corporation, following his resignation from Phillips in Midland, Texas—where he had been working with computers in data management and oil exploration. Initially Andy moved to Denver to establish our office, and later he relocated in Billings. Our early concentration was on a data management consulting project for IIAPCO (Natomas) in Indonesia. In addition, Compex acquired interests to oil and gas leases in the Bradshaw gas field area of Kansas, which my personal company, Remuda Corporation, had previously leased. My first exposure to international deals had been in 1970 when I acquired an option from Kondur Petroleum on an Indonesian concession in the Malacca Straits. This opportunity was introduced to me by Don Todd, a friend who was one of the founders of IIAPCO. I then negotiated a deal with Houston Oils Ltd. and Pan Ocean to acquire the concession, thereby giving Compex a small finders fee of Houston Oils Ltd. stock. This concession, after beginning with disappointing dry holes, later became productive and contains the Lalang field, which was producing 30,000 BOPD in 1984.

After Dan Williams resigned from Phillips as Chief International Geologist in 1971, he joined Andy and me to form the Marinex companies and he moved to London to open an office for Marinex in a flat in Kensington near The Geological Museum. Within a short time, we needed larger quarters and rented a house at 12 Perceval Avenue in the northern London suburb of Hampstead. These quarters gave us abundant working space, as well as living quarters, which sufficed until we temporarily closed the London office in 1978 and Dan moved to Houston, Texas.

During the early Marinex (and Compex) days, we had more ideas than capital, but we were fortunate in forming exploration joint ventures with good partners. Our first financial support from outside our own resources came from Tommy Hendrick, who was a major shareholder in a small exploration company, Hadson-Ohio Oil Co. of Oklahoma City. Hadson's backing in early 1971 was important in covering the expenses we needed to form international exploration groups. The first of these groups was for the North Sea Fourth Round in the U.K. The first exploration joint venture (1971) was with Houston Oils Ltd. (Calgary, Alberta), run by Al Whitehead, and with Cardinal Petroleum Company (Billings, Montana), run by Hugh Palmer. During our 2-year relationship, we acquired licenses in offshore Spain and offshore Netherlands and made several applications in other countries. In 1974, we formed a joint venture for worldwide exploration with Dome Petroleum Ltd after extensive negotiations with top executives. This was short-lived (1 year) because of a change in direction by Dome's Board of Directors. We also formed exploration groups with Natomas, Global Marine, Hadson, DNO, Carless, Bomin, and others. These groups were usually of short duration (1-2 years) depending on success and budgets. Most of these group acquisition programs were in the North Sea (U.K., Netherlands, and Norway) or Spain.

EARLY EXPERIENCES IN ENGLAND

I have fond memories of the 12 Perceval Avenue house and the neighborhood, as it was a stimulating and exciting time. During the latter part of 1971 my wife, Rachael, and I lived in this house for a short time with our sons, Lang and Craig, who were $2^1/_2$ years and 7 months, respectively. Although our life was quite hectic, it was also a rewarding experience.

Some flashbacks:

• During our short residence in London at No. 12 Perceval, with our two young sons, Rachael called a utility serviceman to improve the heating system. He checked the hot water system and informed us that houses in England are not designed to be as hot as in America!

• During a trip to Norway when Marinex was operating a Norwegian North Sea Group, Dan and I could only obtain room reservations on the outskirts of Oslo. The hotel was quite adequate but the fire escapes from our third floor rooms consisted of large hemp ropes dangling to the ground from our windows. This was one time that I noticed the fire exit! Also, on a trip from London to Oslo, I neglected to bring my passport. Although I was able to go in and out of Oslo customs, it was necessary for Rachael to bring my passport to Customs at Heathrow upon my return before I could leave the airport.

• One of our favorite restaurants, the "Welcome Chinese" in Belsize Village, a few blocks from our house, is fondly remembered for its infinite number of courses, each of which was served only after the previous one was consumed. Another neighborhood favorite was "Keats" restaurant, which had an excellent mixture of good food, history, and atmosphere.

• The tradition of serving coffee only after dessert had been finished sometimes irritated us Americans.

• Business efficiency was often frustrated by the slow turnaround of tasks like reproduction and printing. In addition, hard data, such as logs, well information, seismic, etc. were not usually available. Many office buildings were locked on weekends, even to tenants. At times, the long lunches would wipe out most of the afternoon. However, in

future years these lunches became more enjoyable and worthwhile. The cost of everything always seemed high, but a pound would usually buy about the same as a dollar in the U.S., regardless of the exchange rate.
- The rewards of the London area included deep tradition and history combined with unlimited exposure to the best in music, museums, theatre, lectures, and the like. The large and well-maintained parks were oases in the city.
- The "language barrier" between American and English vocabulary was always a generator for a good sense of humor.
- The southern England countryside provided a never-ending source of beauty.

U.K. OFFSHORE FOURTH ROUND GROUP—1971

Following the formation of the Marinex companies, we immediately began plans to form a group to participate in the Fourth Offshore Round in the U.K., an area in which Dan had considerable background. The record of successful oil and gas finds by Phillips in the North Sea was impressive, and Dan's background was the strong drawing card to the group. Another important ingredient in our credibility was that Dudley Tower agreed to serve on the boards of Marinex and Compex. Dudley was a friend with whom I had worked in the mid-1960s as a consultant for T & T Oil Co., which in turn had been on a retainer for Union Pacific Railroad. Dudley had vast corporate and international experience from his tenure with Union Oil, for whom he had served in various capacities up to the position of president. He also loved to visit Montana, where we spent many enjoyable hours, some of which required fly fishing for trout.

Another important element during the early years of Marinex was the strong legal support we had in the person of Stanley Cunningham, who was with the Oklahoma City firm of McAfee, Taft. Stan had considerable international experience in oil and gas transactions and had previously worked for Phillips.

The Fourth Round group was operated by Clinton International in London with Marinex (Dan Williams) acting as exploration advisor. The group consisted of Hadson-Ohio, Clinton Oil Co., Zapata, Bomin, Canadian Export, and Carless Petroleum Ltd. Some of these companies would also join us later in our southern England group. The group was successful in acquiring four attractive blocks in 1971–1972, two in the oil basin, and two in the southern gas basin. Marinex/Williams formed the group and brought Carless-Capel and Leonard, Ltd. into their first oil exploration venture. Carless, with their reputation and history, was an important part of our group and they later became the operator of the Humbly Grove discovery.

The first well to be drilled by the group was in the southern North Sea gas basin (Block 43/11), and Marinex, in order to help finance its interest, sold 5% of Marinex of England Inc. to Westburne Drilling of Calgary. We were assisted in this financing effort by W.W. Charlton from Calgary, who also was quite helpful in subsequent financing efforts. This well, drilled in 1973, was "high and dry" and the anticipated reservoir was salt plugged.

The second well drilled by the group in the North Sea was a different story. Marinex was required to furnish a bank guarantee for their portion of drilling a well on Block 21/2 in December, 1974. In order to help fund this, we were negotiating with three companies on a financing arrangement. During the final sessions, I recall long hours in New York City hotel rooms and lawyers' offices. To expedite the deal we circumvented the lawyers by signing only a short letter agreement with St. Joe Petroleum and Candel. It was a bright day when I flew back to Billings with a "check in fist." This proved expeditious, since the full agreement took years to complete. These negotiations were finalized due to the efforts of two oil field veterans, Chase Ritts with St. Joe and Bill Leuschner with Candel.

The 21/2-1 well discovered the Glenn field in June, 1975 and was tested at the rate of 5600 BOPD from 160 feet of Middle Jurassic sand at 3904 meters (12,809) feet. We subsequently celebrated, considering ourselves "rich and famous" and looking forward to a healthy future income. That was the good news. The "hangover" occurred over the following 15 years, when seven very expensive wells were drilled with only partial success and no cash flow.

THE NORWAY FIRST OFFSHORE ROUND—1972

Marinex formed a group of companies to participate in the first proposed Norwegian offshore round. Natomas acted as the operator and Andy Fish moved to Oslo in 1972 to open an office for our group, which was known as the Odin Group. The Norwegian government had been encouraging all companies, both majors and independents, to participate in the exploration of their portion of the North Sea basin. Our group was strong and we invested much seismic, time, and expense in preparation for the offering. However, after two years of such investment, in the 1972 round only a few blocks were awarded and these only to the major oil companies. We didn't go back to Norway!

THE HUMBLY GROVE GROUP

An exploration program in a southern England onshore basin, in the Jurassic (which was the main source of the North Sea basin oil reserves), appealed to our company, Marinex Petroleum Ltd. This was an expansion of ideas from Dan Williams following an earlier visit to the old Kimmeridge Field during a field trip to southern England. This field produces oil from a thin, fractured Jurassic section.

In 1975 Marinex formed a group of companies to bid for licenses in southern England, with Carless acting as operator and Marinex (Dan Williams) acting as exploration consultant to Carless. Our early discussions with Carless, the operator of our southern England group, were primarily with John T. Leonard, their managing director. He later became chairman, and A. J. Goodfellow became his able fellow executive director. John was a direct descendent of one of the founders of Carless. Although they seemed to like the "revolutionary" idea of participating in an oil and gas exploration effort, it was obviously not an easy task for them to convince their board of directors to commit funds to such a risky business. However, they did join our U.K. Offshore Fourth Round Group in 1971 and the southern England applications in 1976.

Because of their outstanding reputation in the British community, Carless was a large contributor to the success of the early Marinex groups in obtaining licenses. Carless had been founded in 1859, and was undoubtedly the oldest oil company in the U.K. The company originally was engaged in the distillation of coal tar derivatives, and later was recognized as the original refiner and distributor of "petrol," to which they gave this name when introducing it to the British market at the end of the last century. The company refined the by-products of the coal gas industry for several decades, but as natural gas became more accessible, they switched almost entirely to petroleum-based feed stocks. Their refinery in Harwich, built in 1964 on the North Essex seaboard, processed natural gas from the southern North Sea. Carless, Capel & Leonard Ltd. became a listed company on The Stock Exchange (London) in 1971 and their exploration efforts later provided the largest assets of the company.

The areas we applied for in southern England had been sparsely drilled and had had only scattered oil and gas shows. Most of the previous wells had been drilled based on surface structures or shallow reflection seismic. Much of this acreage had been licensed to British Petroleum (BP) in the past and they had later relinquished most of it as being non-prospective.

Needless to say, there were some doubts expressed in forming this

Pump jacks at the Wytch Farm Field.

group since we were second-guessing BP in their own backyard. One of the most interesting individuals we discussed this idea with was Joseph Hirshhorn of the Hirshhorn Gallery in Washington. He liked our idea but probably could not fully appreciate it, just as I could not fully appreciate some of the art work in his Manhattan apartment. The final makeup of the original group consisted of Carless 30%, Marinex 30%, Hadson 20%, and Canadian Export 20%.

In London, from 1974 to 1978, rumors kept surfacing of a potential, big discovery by British Gas/BP in Dorset near the south coast, but at that time even the optimists could not have conceived of the size of the Wytch Farm field. It was even then rumored that BP was reluctant to drill the Wytch Farm discovery well in 1974. The field later (over 15 years) showed reserves in excess of 300 million barrels of oil from the Jurassic and Triassic and is currently being developed to produce 60,000 BOPD or more.

The Marinex/Carless group was awarded five blocks (two licenses) in 1976 and with subsequent acquisitions became, along with Conoco, Amoco, and BP/BGC, one of the most dominant exploration consortiums in southern England.

THE U.K. ONSHORE LICENSING SYSTEM

Determination of full ownership to the mineral rights in oil and gas of onshore Britain was first established under the Petroleum (Production) Act. Earlier, however, the government had conducted successful drilling to the north near Chesterfield on the Chatsworth estate of the Duke of Devonshire, in order to secure oil supplies during the First World War (1914-18) under the Defense of the Realm Act. When this act expired, a settlement was made with the government and the well was taken over by the Duke of Devonshire. This well remained modestly productive until the 1940s. During the 15 years in which the 1918 Act was in force, only seven licenses were granted, three of which remained in effect when the Petroleum (Production) Act of 1934 was completed.

The 1934 Act effectively nationalized mineral ownership of oil and gas in Britain by vesting in the Crown the ownership of all naturally occurring petroleum with the exception of the three previously issued licenses. This act is still the authority under which onshore licenses are issued. The two types of licenses were: (1) Prospecting Licenses, which carried drilling rights and covered 21 to 72 square kilometers (8 to 28 square miles) with an initial term of three years and an option for two extensions of 12 months each; and (2) Mining Licenses, covering 10.36 to 259 square kilometers (4 to 100 square miles), with an initial term of 50 years and the option to extend another 25 years. Royalties to the Crown on oil and gas vary from 5% to 12.5%, depending on production rate. Applicants for the licenses had to provide evidence of financial capability and technical competence. Following this legislation, three principal players obtained licenses: D'Arcy Exploration (later BP) with 18,551 square kilometers (7163 square miles), Anglo-American Oil Co. (later Esso) with 6123 square kilometers (2364 square miles), and Gulf with 3905 square kilometers (1508 square miles).

These license terms were in effect until 1966, when the rights were combined into one form, the Production License (PL), which was tailored to conform with the more modern North Sea licenses. Although environmental issues were noted even in the early drilling days, not until 1972 were major changes made in the license terms. At that time, an exclusive Exploration License (XL) would cover the initial geological/geophysical stage but would not allow drilling. These new provisions for both XLs and PLs required that certain operations, such as drilling, must have planning permission prior to the government's approval. The Petroleum and Submarine Pipeline Act of 1975 tightened some controls over the existing PLs. The aforesaid terms were in effect on most of the Marinex licenses and at Humbly Grove (PL 116).

During 1978, the government required that new license applications offer the state oil company (BNOC) the option of a 51% equity participation, although this was not activated onshore. This requirement was removed in 1980 when the Conservative government was elected. The new government replaced the equity participation option with a BNOC option for a 51% call on the oil at market price.

Onshore fields have been subject to Advanced Petroleum Revenue Tax (APRT) since 1982. This is only in effect, however, if the field produces in excess of 3.7 million barrels per year (10,000+ BOPD). Certain allowances or deductions can offset this tax, but the tax can be as high as 75%. The U.K. corporate tax rate is 35%.

In summary, the U.K. licensing system has been very effective in developing the oil and gas resources in the country. The work commitment program, rather than large bonus bids, generated more drilling and also encouraged smaller companies to enter the play with new ideas. The bureaucratic system, although sometimes slow, usually was fair and reasonable.

SOUTHERN ENGLAND'S GEOLOGICAL BACKGROUND

The depositional basins of southern England are southwest of London and extend south into the English Channel and west to the outcrops in the Exeter region. The principal subbasins are currently known as the Weald, Wessex, and English Channel basins and consist primarily of Mesozoic sediments overlying an old Paleozoic platform. Areal extent of the exploration play in these basins is approximately 241 by 113 kilometers (150 by 70 miles). Triassic and Jurassic reservoirs contain all the oil and gas found to date, except for a small amount of gas found in a Cretaceous reservoir. The important ingredients for petroleum accumulation are the development of the reservoir beds, accessibility to the Jurassic source rocks, and structural configuration.

The Triassic basin is best developed along the south coast around Bournemouth, and it extends northward in a narrow trough to the Gloucester region. The Sherwood sand of the Triassic at Wytch Farm contains the largest oil reserves in the onshore U.K. This fluvial sandstone sequence attains a thickness up to 1000 feet. The Jurassic strata that have provided the source beds for the oil and gas contain two principal pay zones, the lower Bridport sand and the middle-Great Oolite limestone. The Jurassic sequence contains a thick series of marine shales with some limestone and sandstone, and is widespread across much of southern England. During the Middle Jurassic, a thick sequence of oolitic limestones was deposited in the Weald basin and the northern part of the Wessex basin. This sequence, known as the Great Oolite, is the main pay at Humbly Grove, which is situated on the north flank of the Weald basin.

Humbly Grove #1 well in a typical English countryside setting.

During the Late Jurassic, major growth occurred on faults bounding these basins, and additional deposition of massive marine shales followed. After the end of the Jurassic, uplift and erosion occurred around the margins of the basin prior to deposition of a continental Cretaceous sequence. During the Late Cretaceous, a thick and extensive chalk was deposited. At the end of the Cretaceous, Alpine compressional movements gradually inverted the Weald and English Channel basins. This inversion, which culminated in the Miocene (Tertiary), has complicated geological interpretation since the last century, and only in the late 1970s, with the aid of modern seismic data, did it become apparent that the surface geology bears little relationship to the subsurface geology. Petroleum accumulation in subsurface tilted fault blocks is either offset from surface features or does not even relate to them. Much of the early drilling had been based on large surface anticlines and was unsuccessful. The current depth to the base of the Mesozoic in the southern England exploratory area is usually in the 1550–2790-meter (5000–9000-foot) range.

HUMBLY GROVE PROSPECT

Following extensive seismic programs on the Carless Group's licenses, various prospects were identified and locations selected based on geology and license requirements. The first license selected

Our "well-sitting" pub.

for drilling was PL 116, south of Basingstoke in Hampshire County. This drill site was chosen based on the seismic interpretation of a Middle Jurassic structural closure covering approximately 3.72 square kilometers (920 acres) and having the potential of up to 7.69 square kilometers (1900 acres) depending on the extent of the closure to the east. The feature was bounded by normal faults downthrown to the north and south. The drill site was in a "pea patch," as it was described later in newspaper accounts, west and next to Highway A32. An important factor in drilling operations was the approval, by various local agencies, of planning permission to drill a well. This was, and still is, particularly difficult, especially in areas of extreme beauty where local people have little or no background in oil field operations and where the landowner has no mineral rights. It might be necessary to contact twenty or more agencies in this process. Two of the most important approvals that must be obtained prior to drilling are those of the landowner and the County Council. Carless demonstrated their willingness to work with the local public in the drilling areas by relocating an exploratory well away from a site on some scrubland that was the home of an extremely rare colony of Duke of Burgundy butterflies.

The Humbly Grove #1 well was spudded in April, 1980, with Dan

Williams acting as the well-site geologist. Progress was slow because of extensive coring in the Jurassic. During the drilling operation, I flew to England from Billings, Montana to visit Dan because he had been on the well for quite some time. Surely, I thought, he was growing weary of the "well sitting" routine of dingy motels, remote areas, and bad food that we were accustomed to in the U.S. Upon my arrival at Gatwick airport, we drove immediately to the rig, reviewed drilling operations, and then drove a few miles to have lunch. During lunch I realized that this was not typical "well sitting" duty. We ate an excellent, traditional pub lunch under a 200–300-year-old tree in a sunny courtyard adjoining a 17th-century graveyard. This small, quiet village was situated on one of a group of rolling hills in the beautiful, lush, green countryside. Following lunch, we went back to the rig for a few hours, and then drove into Alton, the home of Jane Austen, where we stayed at the Swan Hotel, a picturesque old English village hotel. This was a good duty!

Oil staining was cored in the Great Oolite, and after a failed drill-stem test and much debate, production casing was run. The well encountered up to 41 meters (136 feet) of oil zone in the Great Oolite formation of the Middle Jurassic at a depth of 1173 meters (3850 feet). Short production tests indicated that the well was apparently commercial, with a rate of 72 BOPD of 39° API high-quality sweet crude. Total time for drilling and testing had been 55 days.

POST-DISCOVERY—THE "LAUNCHING" OF MARINEX

Following the elation of the discovery we began to discuss further plans for our newly successful company. Dan had suggested, during the drilling of the well, that we consider going public with Marinex; I had previously resisted operating within a public company. However, after considering the timing and circumstances, we agreed to attempt a public offering in the U.K., and Dan consented to run the company. Since I had little background in this arena, I called my friend Chuck Charlton in Canada to assist us. He was familiar with our company through his previous assistance, and he had considerable background in public offerings as well as the workings of the financial community. The launching of Marinex was the ultimate in riding the crest of the oil boom. The "hype" was intense in the British press and rumors were abundant. The favorite television show of the day was "Dallas," and the press expanded on this and speculated about various "J.R.s." We presented our company, with its interest in many attractive prospects in southern England in our land position of 5259 square kilometers

(1.3 million acres). The first of these prospects that was drilled had resulted in the Humbly Grove discovery a month earlier. In addition, Marinex was to have no cost in the drilling phase until a commercial well was encountered on each license. This was during the time when oil was predicted to be $50.00 per barrel and demand was expected to increase "forever." We essentially had our choice of underwriters, brokers, merchant bankers, and the like, and also selected some of the best lawyers and accountants available. Our selection of brokers/underwriters was made based upon the best trade with the company having the strongest, most aggressive approach to "do the deal." After extensive negotiations, this proved to be Carr-Sebag, who had a long history in the London financial community. (Carr-Sebag later merged into what is now the Kleinwort-Benson Group.)

I recall a meeting in the lawyers' conference room near St. Paul's Cathedral in June 1980, when it seemed as though hundreds of "meters" were ticking at a huge table surrounded by our advisors. Our down side of costs would be considerable if the offering were unsuccessful. We finally agreed with Carr-Sebag on terms wherein the three of us retained some cash and we also, along with Chuck Charlton, retained much of the company. The public offering raised £8.64 million of equity capital intended for the development of the Humbly Grove field and other ventures.

We finalized the documentation and negotiations in record time and from the start-up in early June, 1980 we began trading on July 22, 1980. Chuck Charlton was pushing us to complete the transaction as he anticipated a downslide in the market. This proved to be correct as our agreed price of £1.60/share dropped to approximately £1.30/share almost immediately but recovered to about £1.70/share before dropping further as the recession began.

DEVELOPMENT OF THE HUMBLY GROVE FIELD

The discovery well at Humbly Grove tested oil in May, 1980, but it was shut in until December 1980. The oil was a high-quality (39° API) crude that resembled the crude at Wytch Farm. The Humbly Grove discovery was the first ever in Hampshire County, which is one of southern England's most beautiful areas. The development of the field was painfully slow by North American standards, mainly because of the requirements and approvals of various governmental agencies. Environmental concerns were addressed and the education of the local authorities on oil field operations was required. After a few months, the well stabilized at a rate of 100 BOPD. Following further seismic surveys and the granting of planning approval, three appraisal wells were drilled from two new sites in February, June, and August of 1982. These three wells revealed the gas cap and the water leg, as well as the presence of a lower zone in the Great Oolite that contained low permeability. In addition, the #2 well revealed a new pay zone in southern England, the Rhaetic formation at the base of the Jurassic. Following this successful appraisal program, the proven and probable reserves in the Great Oolite limestone were estimated to be 13 million barrels of oil and 3 bcf of gas cap.

A planning application for the development of the field was submitted to the Hampshire County Council in February 1984 and was approved in July 1984. The Annex B (development program) was submitted to the Department of Energy in July 1984 and was approved by the Secretary of State for Energy in January 1985. This delay of almost 5 years in development was more comparable to a North Sea discovery than an onshore find and caused considerable cash flow problems for Marinex and other partners in the group.

The approval of the Annex B development program was based on a Phase 1 program to develop the upper, high-permeability reservoir and to appraise the lower, low-permeability reservoir and the deeper Rhaetic formation. Thirteen development wells were drilled between May 1985 and May 1986 from three surface locations, and twelve of these wells were deviated to minimize the environmental effect of surface drill sites. This had also been the common practice in the Wytch Farm field, 60 miles to the southwest in Dorset County. During the initial development phase, a gathering station was installed together with a 7.2-kilometer (4.5-mile) pipeline that moved the oil to a rail export terminal where it could be hauled to the Esso-Fawley refinery, 130 kilometers (50 miles) to the south, on the coast near Southampton. The Humbly Grove oil field was officially commissioned with much fanfare on June 4, 1986.

The field equipment was designed for an expected 4000 BOPD and a GOR of 2000. Subsequent production history averaged 1000–1500 BOPD and the GOR was much higher, especially in the upper zone. The gas produced in the early years was reinjected into the reservoir at enormous expense, due to the electric power cost for the recompression units. The options for selling this gas were limited because of the small quantity available and the difficulty in dealing with only one potential buyer, British Gas. However, following the Energy Act of 1983, gas could be converted and sold into the national electric grid. The Carless Group then decided to install a gas turbine generator to meet the field power requirements and export the surplus power to Southern Electricity. Another generator may be installed in the future.

The field had produced almost 2 million barrels of oil and 1.5 bcf of gas by the end of 1989. Production at that time was about 1000 BOPD and 5.5 MMCF per day. The Phase II development program has been delayed (possibly forever), but the economic life of the field has been extended because of significantly reduced operating costs as a result of the gas sales scheme.

The Humbly Grove field was developed by adopting high standards of environmental protection and by employing optimally designed equipment. To a great extent, this field served as the "guinea pig" for future oil field developments in southern England. The working interest owners of the field were economically penalized by this learning curve, but the successful results of the development would pave the way for future fields that could be developed economically and with minimal environmental impact.

EPILOG

The history of the Humbly Grove discovery is one of many stories of the worldwide search for oil. It is a somewhat unusual tale, however, in that during the short span of 9 years, three individuals started a private company with limited funds, and the company went on to be capitalized in a public offering for £43 million (approximately $100 million). This occurred because of the fortuitous combination of the timing of early exploration success on a large land position, the peak of the oil market, and the people involved within an arena of free enterprise. The Carless Group made four discoveries out of the first five exploratory wells drilled in southern England, which certainly justified the early optimism. Some of these had substantial oil in place, but good recoveries have continued to be elusive in the Great Oolite reservoirs. Many people think that Wytch Farm is a "freak" and that there are no other comparable fields in this area, but can there only be one major oil field in this sparsely drilled region?

Following the recent downturn of the petroleum industry, which began in the early 1980s, the companies in the initial Humbly Grove discovery well have virtually ceased to exist with regard to their names and original ownership. The control of Marinex was sold in 1983 to obtain financing, and was later merged with Teredo. Hadson sold their British subsidiary to Britoil; St. Joe ultimately became a part of Petrofina and more recently Monument (in the U.K.); Candel was absorbed by BP, and even Carless, which had survived for 129 years as a family-controlled business, was acquired in a hostile takeover by Kelt in 1988.

There is a continuing, positive result of our group's exploration effort in the U.K., however. Secure onshore domestic oil reserves are available. In addition, they were developed in a region of outstanding beauty, without adverse effects on the environment.

Interior Sudan, 1973–1980

Beyond Khartoum: Petroleum Exploration and Discovery

By Allan V. Martini
with
James L. Payne

When I joined Chevron Overseas Petroleum Inc. as Vice President–Exploration in early 1973, the company was actively exploring for oil in east Africa, with programs in Madagascar, Ethiopia, and Kenya, and a budding interest in Egypt. It was a time when new technologies and concepts, particularly organic geochemistry and plate tectonics, and new tools, such as satellite imagery, were beginning to be actively applied in petroleum exploration. It was also a time when declining oil production in the United States and increasing demands by the OPEC countries were leading to rapidly expanding foreign exploration activity in a number of new areas.

By 1973, oil production had been established across North Africa, from Egypt to Algeria and down the west coast from Nigeria to Angola. However, the industry's activities tended to be confined to the margins of the continent. Most of the interior, in addition to being very remote, was generally considered to consist of exposed "basement type" igneous and metamorphic rocks that were covered, in occasional shallow, intracratonic sags, by a few thousand feet of Cretaceous, Tertiary, and Quaternary continental sediments.

AN OBSERVATION— AN IDEA

In February 1973, John Miller, a geologist on Chevron's San Francisco staff, wrote a memo to W.M. Chapman, Resident Manager of Chevron Oil Company of Kenya in Nairobi. Miller was making a regional geologic study of the Kenya concession using earth satellite images and what little surface and subsurface geologic data were available. Miller noted:

> In working with the Kenya geology, it was apparent to me that there may be a structural trough or graben extending from the southern part of the Sudan into Kenya...Lake Rudolph lies precisely between what I imagine to be two boundary faults. The Bahr el Arab drainage and the Bahr el Jebel leg of the Nile River mark its trend in the Sudan. Other indicating features [include]...an anomaly in the pattern of the "Rift Valley" volcanics...where the rift valley crosses the postulated trend...the volcanic belt flares out southeastward toward our concession area, and northwestward toward the Sudan.

Over the next several months, Chevron geologists expanded Miller's observation into the concept that the Sudan depression might be the surface expression of part of an aborted interior African rift system. The rift could be postulated to extend from the Benue trough in Nigeria, through Chad into the Sudan, and southeastward into the Lamu embayment in Kenya. It would have begun to develop in Late Jurassic–Early Cretaceous time, coincident with the breakup of Gondwana and the separation of Africa and South America.

Chevron and Texaco had an agreement providing for joint exploration activities in several countries in east Africa and Asia, including the Sudan, and in late 1973 we found that Texaco had developed a similar concept. Geologists from both companies noted the presence of marine Jurassic and Cretaceous sediments in Kenya and Ethiopia and speculated as to whether such sediments could be present in significant amounts in the Sudan.

Our growing interest in the Sudan was further stimulated from another direction when the Egyptian government announced that it would consider exploration proposals on unleased areas in the southeastern portion of the Gulf of Suez, where recent discoveries by Amoco had aroused the interest of the industry. A joint Chevron-Texaco team met in Cairo to examine technical data, and an exploration proposal was submitted to the Egyptian government. That first proposal was not successful, but during its preparation the possible extension of the Gulf of Suez geology to the south along the Red Sea coast of Egypt and the Sudan became apparent. Our records indicated that AGIP had drilled several dry holes on the Red Sea coast during the 1960s, but we had no details regarding the results. Commercial scouting services reported that several small independent companies held exploration permits in the Red Sea offshore, but it appeared that most of the area was still unleased.

FIRST CONTACT

By February 1974, it was clear that someone should visit Khartoum to look for geological data in government files from both the interior and Red Sea areas, and to assess the lease situation, petroleum regulations, and general conditions in the Sudan. Bill Chapman, manager of Chevron Kenya in Nairobi, and the recipient of the original Miller memo, was a geologist with many years of experience in foreign exploration and in dealing with foreign governments. He agreed to make the visit.

Chapman spent a fruitful six days in Khartoum, and on March 28 forwarded a report to D.O. Nelson, President of Chevron Overseas in San Francisco, which read in part:

Nairobi
28th March, 1974
D.O. Nelson/For J.M. O'Connor
and A.V. Martini

The following discussion covers my recent visit to Khartoum over the period from March 20th through March 25th. I departed Khartoum via Addis for Nairobi on March 26th....

Most of the stay in Khartoum was spent at the Mining and Geological offices under the auspices of the Ministry of Industry and Mining. Shortly after arriving in Khartoum on Wednesday, I called at the office of the Director for Geological and Mineral Resources. The director's name is Abdel Latif Widatalla. After a visit with the director, he arranged for me to return on the following day to the mining department, where all of the information in their files relevant to petroleum exploration in the Sudan would be at my disposal. I had given the director a copy of the Socal Annual Report, explaining that we wished to make a geological review of his country for the purpose of determining its petroleum potential. I also suggested that if the review were favorable we would very likely wish to present an application for an oil exploration permit. During my initial visit, the director introduced me to his assistant petroleum engineer, Dr. Omar El Sheikh. Dr. Sheikh was instructed to give me any assistance I required....

I was impressed by the level of education and technical understanding demonstrated by the director and his assistant, Yousif Sulieman, as well as several technical people I met with in the Mines Department. Most of these people have been educated at Khartoum University and the University of Moscow. The director was trained in England.

Since all laws and regulations pertinent to oil exploration are attached, I will not herein discuss them in any detail. The government's attention towards petroleum exploration investment is quite favorable, which is understandable because of the lack of past exploration's success and what may be a relatively low level of industry interest. All present existing license areas are in the Red Sea area. They are held by American Pacific, Ball and Collins, and Oceanic Exploration. Adobe's license in the same area, will, I understand, be executed shortly. I do not know their precise area of interest. The director advises me that if we wished to obtain a license in the interior of the Sudan, appreciably better terms (than those given in the Red Sea) can be negotiated.

The government wants a 50/50 division of profit and is prepared to negotiate such things as tax and to the extent where they would not interfere with this profit division. Suitable investment guarantees, involving repatriation of investment and profits in hard currency, would be no problem, according to the director. Exemption from duty and tax for all exploration equipment, material and supplies is provided in the regulations. The regulations state that oil mining permits require incorporation by the permittee in Sudan. This requirement can also probably be removed through negotiations. Although established procedures do not provide for the attachment of oil prospecting and mining permit terms to the OEP, these terms can be worked out while negotiating the oil exploration programme. Apparently, existing licensees have not negotiated these two latter conditions.

While accumulating the attached material, I did not find time to make a geological evaluation, and, I did not feel that such an evaluation was expected. However, the AGIP final relinquishment report is based on six wells and extensive geophysical work. Upon completion of this work, AGIP did not regard the Sudanese Red Sea area as geologically favorable. Two areas in the interior have suggested some possible potential. The first, in the northwestern Sudan, was evaluated by Continental with an aerial magnetic survey. Results of this survey are attached and indicate that the sedimentary section is thin and spotty over most of the area, with westward thickening zones near the Sudan-Chad border. The sedimentary section is Paleozoic, as evidenced by the Paleozoic outcrops along the border. The second potential area appears to be a northwest-trending regional graben lying across the southern Nile drainage. The area is seismically active, which suggests continuing subsidence of the graben and possibly some transcurrent adjustments. The graben trend appears to project very neatly into the Lake Rudolph graben. An F.A.O./U.N. geophysical survey confirms the presence of a graben and at least a moderate thickness of sediments. Horst-graben structures are also suggested by results from this survey. The survey was done by Hunting Geophysical in the summer of 1973 and it largely consisted of gravity work. Topography also confirms the presence of the graben, with relief changes of about 500 meters from scarp to scarp across the graben. The graben is also the site of a huge papyrus swamp where the Nile river flows across the graben. You will probably recall that this swamp served as a major hindrance to the efforts of several expeditions to locate the source of the Nile. Another point of interest is that much of this portion of the southern Sudan, because of its remoteness, is considered to be one of the best remaining game areas in Africa.

Khartoum is very interesting. Its residents make up a representative cross section of the various ethnic groups in the Sudan. Most are Moslem, with Egyptian Coptic Christians, and animist tribes. The people are handsome, well-built, and Arabic in behavior. The climate of Khartoum is usually quite hot—with the exception of two or three months each year when it is comfortable. During my stay, the temperature was moderate although the hot season had supposedly started…. The Sudanese are strongly allied with Egypt and Saudi Arabia but maintain cool relations with Libya. The Russian influence has been reduced since 1971. The American Embassy appears to have a great deal of confidence in the present Sudanese government…and states that relations between the U.S. and Sudan are good…and that foreign investment, particularly American, will be more secure in the Sudan than many other African and Arabic countries…. The Democratic Republic of the Sudan is ruled by the military, who appear to have pulled the various dissenting tribal groups into a peaceful, if not completely harmonious, association. President (Major General) Nimeiri has brought a new constitution into being and has made provision for a legislative assembly.…

In summary, my visit to the Sudan yielded positive impressions prompted by such things as the friendly, cooperative attitude of government officers and the general public. There seems to be a prevalent friendly, and possibly even pro-American, attitude. The professed government policies towards private foreign investment are cooperative and encouraging.

W.M. Chapman

Certainly the most exciting technical find of the visit was the geophysical survey mentioned in Chapman's report, which was part of a ground-water study carried out for the Food and Agricultural Organization of the United Nations. The survey consisted of a series of gravity meter and refraction seismic profiles obtained in the vicinity of Babanusa, a small railroad junction located in the area of the suspected interior sedimentary basin. The data were reviewed in San Francisco by Fred Flege, a gravity-magnetics specialist assigned to work with Chevron's Africa Exploration Area. He concluded:

- Major gravity minima of 40 to 50 milligals amplitude are present on lines 2 and 3.
- Minimum sediment thickness is in the gravity minima range from 2500 to 3000 meters.
- Gravity suggests the presence of major faults in the area.
- Estimates of section thickness are questionable because of lack of knowledge concerning sediment densities, basement compositional effects, and the nature of the regional gravity field.

While far from conclusive, this was the first direct evidence that the postulated sedimentary basin might, in fact, exist.

The legal basis for oil exploration and production activity in the country was the Petroleum Resources Act of 1972. It established a system of Exploration Licenses convertible to Oil Mining Leases following discovery, which was similar to systems we had encountered in several other former British colonies, including Kenya and Nigeria.

While the Act was broadly satisfactory, there were some practical problems:

- There was no limit on the number of licenses that could be held, but an exploration license could not exceed 800 square kilometers (about 300 square miles).
- The maximum term of a license was four years plus a two-year renewal, for a total of six years. Although this was reasonable for a readily accessible area such as the Red Sea, it was completely impractical in the interior.
- Most importantly, there was no standard form of Oil Mining

Lease. While the Act treated at length the royalties, government participation, terms of lease, special taxes, exemption from tax, and the like, there was no way to know exactly how these terms would apply to the successful explorer unless they were agreed upon in advance and attached to the exploration license. Chapman's report had noted, "These terms can be worked out while negotiating the oil exploration program. Apparently, existing licensees have not negotiated these...latter conditions."

Chapman had found a friendly, helpful government, minimal competition, and the promise of reasonable economic terms. While there might be some problems, the huge, unexplored area in the interior, the low level of technical knowledge and the possible presence of a previously unknown sedimentary basin fired the imagination of Chevron's more optimistic explorationists.

ACQUISITION

Bob Daniel, the Africa Area Exploration Manager, and his team moved ahead with preparations to file for exploration licenses. Even before Chapman's visit was concluded, Daniel and his team had contacted the companies that held licenses in the Red Sea, offering proposals that Chevron earn a portion of their interest by assuming their work obligations.

An exploration program in the Red Sea would be the sort of thing we at Chevron had done, and were doing, in many countries around the world. Basically, it would begin with a modern marine seismic survey, followed, if reasonable structures were found, by drilling using a mobile offshore drilling rig. The ease and speed of movement of seismic vessels and drilling rigs in the offshore environment made the planning and timing of this type of exploration relatively straightforward.

The interior area was another matter. While the localized U.N. gravity survey now made the presence of some sort of sedimentary basin seem more likely, we still had little idea of its possible extent. The original projection of the Rudolph Trough of Kenya into the Sudan was based largely on the presence of a topographically low area covered with young, surficial deposits and flanked by Precambrian outcrops. There was no way to know where or how much of this area was underlain by a sedimentary basin. It measured about 1100 kilometers long and from 300 kilometers to as much as 650 kilometers wide. The total area was just over 500,000 square kilometers, which is about the same size as the area of Spain or 80 percent of the area of Texas. We

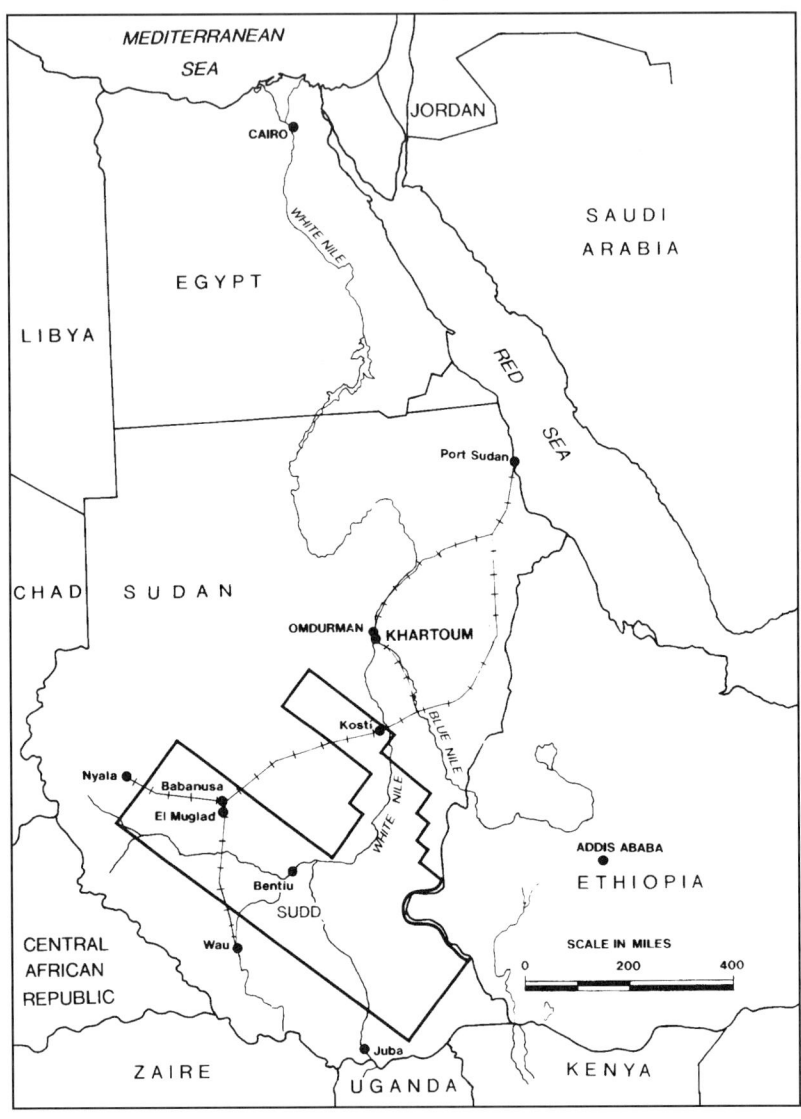

Index map of northeastern Africa, showing the location of the interior Sudan exploration area.

expected we might have some difficulty with the government if we applied for such a large area; at the same time, if we were to assume the risk and expense of determining whether the basin even existed, we had to be sure that, if found, the basin would underlie our licenses.

On May 6, 1974, about a month after Chapman's return, Chevron Overseas submitted a proposal to the Executive Committee of its parent, Standard Oil Company of California. The proposal recommended earning an interest in 13,500 square kilometers held by a small British company, Ball and Collins, in the Red Sea, and filing exploration applications on an additional 31,000 square kilometers in the Red Sea and 516,000 square kilometers in the interior. The proposal concluded:

> In summary we believe the areas proposed for acquisition in the Sudan have major potential. In view of the favorable government terms and improving political climate in the area, we consider an aggressive land acquisition program, as outlined, to be fully warranted.

The proposal was approved on May 9th, Chapman was notified, and he prepared to return to Khartoum. Because Chapman was scheduled to complete his tour of duty in Nairobi that summer, he was accompanied to Khartoum by G.W. Fuller, his replacement as Manager of Chevron Kenya. Gerry Fuller was a geologist whose background in foreign exploration was similar to Chapman's. Fuller took with him from San Francisco the approved range of terms that could be used in negotiating the Exploration Permit applications. In the summer of 1974, telephone communication by satellite had not yet reached the Sudan. An overseas call from Khartoum could take days to go through, only to be unintelligible when completed. The solution, when a call was important enough, was to fly to Addis Ababa or Nairobi to make the call. On May 17, Chapman called San Francisco from Addis Ababa. He spoke with me and John O'Connor, Vice President-Government Relations. O'Connor's notes of the conversation included the following:

- Initial contact with Government indicates we have a good chance of acquiring what we want…this may move pretty fast.
- Apparently inspired by Government, there was a news release in Khartoum yesterday that Chevron representatives (largest oil company in the world, etc.) were on the scene negotiating with Government. Consequently, there may be something in the press elsewhere on this and the competition may be alerted as well.
- With respect to the Interior Block…Government is obviously taken aback by the large size; however, there appear to be no competitors. Government indicated they will want rapid relinquishment and a pattern is developing in discussions toward 25% at the end of two years and 25% at the end of four years.

- It does appear we are facing the heavy volume of concession documentation created by the area limitation on individual licenses. While the Government will tolerate a single reference map for the purpose of the application phase, we will probably have to come up with individual maps for the many licenses. Our reps. will look into this further and let us know what help, if any, they will need from San Francisco.
- Our reps. also told Government that we are obliged to offer sharing of any acreage with Texaco and this door has been kept open.

The applications were finally filed on May 28th. Our proposal for the interior block was for minimum expenditures for geological and geophysical work and exploratory drilling of $1,000,000 in the first year and $2,000,000 each year for years two through six. Chevron would have the option to "back-away" and relinquish all rights at the end of year two if nothing encouraging had been found. If we elected to continue, 25% of the area would be relinquished at that time and an additional 12.5% at the end of years four and six. We would commence drilling the first exploratory well before the end of year four. The proposal included a request for government assurance that licenses still existing at the end of the sixth year would be reissued in the form of new exploration licenses. Chapman's report on the filing concluded:

> The government made no secret of their satisfaction in having Standard Oil Company of California commencing exploration in the Sudan. Three separate press releases were issued by the Ministry of Information during our stay, which announced that negotiations were under way between Ministry of Industry and Socal.

We were elated that things were going so well, but our high spirits got a jolt in early July when the government requested further negotiations. Chapman went back to Khartoum. It developed that Esso, and possibly others, had filed competing applications over all or part of the area. Our concern regarding the publicity surrounding the negotiations in May had apparently been justified.

The government requested that we submit a proposal that would cover only half the area. While on the surface this seemed a reasonable request, we realized it could result in disaster. Given the uncertainty of the location of the basin, if it existed at all, having only half the area could leave us with no part of a new basin after our activity had promoted its discovery. We had to have a chance for at least a quick look at all of the area.

Our response was to submit a new proposal on July 21 offering a $3,000,000 signature bonus and doubling the firm work program, before back-away, if we were awarded all of the area. We also proposed to accelerate relinquishment by surrendering 25% of the area (some 130,000 square kilometers) at the end of the first year and an additional 12.5% at the end of each of years two and three. Finally, we indicated we would consider participation with other companies if awarded the total area. In other words, if necessary, we preferred a half interest in all of the area to a full interest in half the area.

We dealt with the government's request to submit a proposal covering only half the area by "peeling" the outside edges of the original block all the way around, keeping only the central portions farthest from the outcropping basement rocks. It seemed to be the strategy which gave us the best chance under the circumstances. We later learned from government that Esso had received the same request and presented almost exactly the same solution. This convinced some of the more suspicious in government that there must be collusion between the companies, even though the purpose could not be understood!

Although the interior area negotiations were in some difficulty, the Red Sea applications were moving well. The only significant problem that remained unresolved was the lease form. We decided that with indications of growing competition, we should concentrate on acquisition and trust that we could solve the lease form question at a later date. Many months later, we realized that this decision was a key step in dealing with the competition for the interior block. On August 17, Chapman wrote a letter to the government:

17 August, 1974

Director,
Geological and Mineral
    Resources Department,
Ministry of Industry and
    Mining,
Khartoum, Democratic Republic of the Sudan

Dear Sir,

Oil Mining Lease Form

Further to our recent conversations concerning the above subject, Chevron respectfully requests that your Ministry confirm its willingness and intention to develop, at your earliest convenience and during the basic term of the Chevron Red Sea Exploration Licenses, an Oil Mining Lease Form (Contract) which will be mutually agreeable to your Government and Chevron. Chevron does not expect, and is not requesting, that such a Form be developed prior to execution of the exploration licenses or undertaking of the exploration work; however, I believe your government appreciates the importance of having the contract terms and conditions known and agreed upon in advance of an oil discovery.

Chevron will be happy to enter into discussions and provide assistance or cooperation, in any manner convenient to government, in developing the new lease terms. We are ready to provide any information or material concerning forms used by producing countries throughout the world—and to furnish clarification and explanation thereof if desired.

Thank you for your kind attention. I would appreciate your signature signifying your agreement of intent as set forth in the foregoing.

Faithfully Yours,
W.M. Chapman
Chevron Oil Company of Sudan

We heard no more from the Sudan until mid-September, when Fuller transmitted by telex the text of a letter dated September 12. The letter referred to our application for 645 oil exploration licenses in south central Sudan and advised that the application had been "approved in principle." It continued with an invitation to send a representative to finalize the work program and other details and closed with the following paragraphs:

> I also wish to convey to you that the Petroleum Affairs Board has considered the request made in page 6 of your application, dated 23 May 1974, that they exchange letters with you concerning the reissue of licenses, to cover areas held by you at the end of the sixth year of the term of the license.
>
> The Board indicates that they will give sympathetic consideration to an application made by your company for such reissue of licenses at the appropriate time, in the light of the position, progress and results of exploration at the time.

It was clear that the exploration risks in the Sudan would not be confined to the geological risk of locating an oil-productive sedimentary basin. What was the probability we would be successful in negotiating an extension of the exploration permits and a satisfactory lease form?

Fuller returned to Khartoum to conclude the negotiations. Government's primary goal was to further accelerate the exploration program. It was explained that one of the competitive proposals included commencement of exploratory drilling prior to the end of the 30th month, and the government felt Chevron should be able to undertake a comparable commitment. We agreed to commence the first exploratory well prior to the end of year three, and to an increase in several of the minimum expenditure commitments. On September 30, Fuller notified us, "Terms and conditions of all annexes now agreed upon and Government prepared execute licenses as soon as documents available."

"As soon as documents available" brought home John O'Connor's earlier concern over the volume of concession documentation created by the area limitation on individual licenses. Government regulation required that seven signed copies of each of the 645 licenses be submitted. Each copy was to have a plat attached (also signed) that showed the coordinates of the individual license. Computer Assisted Drafting was in its infancy, but it was pressed into service to prepare the plats. The necessary documents, weighing a total of 450 kilograms (990 lbs), were completed in San Francisco and air-freighted to Khartoum where, on November 23, 1974, they were signed in a night-long marathon.

A VISIT TO THE SUDAN

With a major new exploration venture about to begin, we felt it was time for Chevron Overseas management to take a closer look. D.O. "Swede" Nelson, President of Chevron Overseas; Jim Brooks, who had recently replaced John O'Connor as Vice President-Government Relations; and I set out in late November on a visit to Egypt and the Sudan. We were joined in early December in Khartoum by Fuller and John Sutherland, the new Resident Manager of Chevron Oil Company of Sudan.

Our time in Khartoum was filled with the ceremonial signing of a representative license by Nelson and the Minister of Industry and Mining, a reception to meet the other government ministers, a courtesy call on President Nimeiri, and an evening program of Sudanese folk dances. The welcome was warm and enthusiastic. Our only problem was in explaining to our impatient hosts how it could possibly take more than six or eight months before we would know whether or not the oil was there.

Meeting with members of the High Executive Council of the Southern Region, Juba, December 1974. Front row: John Sutherland, Daniel Matthews, Al Martini, Isaac Bior. Back row: Mading de Garang, Hilary Logali, Swede Nelson, Lawrence Wolwol, Abdel Latif Widatalla, Gerry Fuller.

On the morning of December 5th, we departed Khartoum by charter aircraft to visit Juba, 1200 kilometers to the south. Juba was the capital of the Southern Region, and fully two-thirds of the new concession area fell within its bounds. We felt we needed to meet the members of the local government.

The Southern Region consisted of the three southern provinces of the Sudan, and had come into being two years before with the signing of the Addis Ababa agreement. This agreement had ended a 17-year civil war in the south that had broken out in 1955 as the British prepared to withdraw from the Sudan and turn over power to a predominantly northern government in Khartoum. The agreement established the basis for "granting the Southern Provinces of the Sudan Regional Self-Government within a united socialist Sudan." The governing body was the "High Executive Council," led by a president who was also a vice president of the Sudan.

On the evening of our arrival in Juba, we were invited to "a dancing," where we sat outside in the warm, moist, subtropical air, listened to music from a phonograph, and met, informally, some of the members of the "High Executive Council." Many of them had fought in the civil war, and I got the feeling that, perhaps, the war had ended as much from weariness and fatigue as from the resolution of any issues.

The following morning, we had an audience with the President of the High Executive Council, Abel Alier, a graduate of Yale, and a very warm and gentle man. This was followed by a meeting with his ministers. The atmosphere could best be described as polite but guarded. It was soon obvious that they knew an agreement had been made in Khartoum but they knew little else about the matter. We spent the morning describing our plans and answering their questions. When we departed Juba in the early afternoon, we were glad we had made the journey. It had set the stage for good relations between Chevron and the Southern Regional government for many years to come.

The flight back to Khartoum was made at an altitude of about 3000 meters, and for more than four hours, from horizon to horizon, over river, swamp, bush, and desert, we looked down on the contract area. The exploration challenge we faced was impressive.

PARTNERS?

On our return to San Francisco, we received a shock. Texaco, which had joined us for a 50% interest in the Red Sea applications and farm-ins, advised they did not wish to join Chevron in the interior. Without a partner, a venture of this size and risk would be unusual in the international exploration arena even among the majors. While we were digesting the Texaco news, we had a telephone call from Esso Exploration that was confirmed by a letter reading, in part:

> Please refer to our phone conversations...wherein I inquired if Chevron was interested in a partner in the Sud Basin acreage recently acquired by Chevron in the Sudan. Esso...is interested in the acreage and in entering discussions...on the terms under which Esso could participate in the venture should Chevron decide to take a partner.
>
> As I further understand, a Petroleum License has been signed but no form of Mining Lease has been agreed with the Sudanese Government. In our reviews of the Sudan Petroleum Resources Act, 1972, and the regulations thereunder, we found the same lacking in certain provisions and somewhat vague on such matters as the treating of the many Petroleum Licenses as a single unit for a work commitment, the conversion to Mining Leases, and the terms under which a commercial discovery would be developed and produced. The absence of an agreed form of Min-

ing Lease creates a serious obstacle to any Esso participation in the Sud Basin acreage. Esso management would insist on a complete agreement with the government which is mutually satisfactory to all parties before considering any form of participation.

It was now quite clear why the competition for the interior block had seemed to fade away in September. Our letter of mid-August offering to write the Oil Mining Lease later had paid off!

EXPLORATION BEGINS

While the negotiations were going on through the summer, development of the exploration plan had been well under way. Looking for a new basin was certainly starting exploration at square one! Greg Stanbro, the Project Manager, his assistant, Tom Schull, and the other geologists and geophysicists working on the project in San Francisco began to realize they had a unique opportunity to carry out a complete exploration program as one might be described in a textbook—something that almost never happens in fact. In most areas of the world, and especially in the United States, where many of the geologists had been trained, a new program is highly influenced by data from previous efforts; usually by a compilation from someone else's efforts. Any program is essentially a continuation, a repeat, or a fill-in of a previous one. Only a handful of the tools in the exploration kit are applied to the problem at any one time. Here in the interior of Sudan, the immense size of the area, the almost complete absence of information about the subsurface, and the very short time available for evaluation, presented an unusual challenge; it also presented the opportunity to use all the tools in the correct sequence to obtain the best evaluation, in the shortest time, and at the least cost.

The strategy was to begin with methods that covered broad areas at low cost and to use the data obtained to progressively focus the use of more expensive and definitive tools on the most attractive areas. Study of satellite and aerial photos, one of the least expensive reconnaissance methods, had led to the original observation that a sedimentary basin might exist in the interior of the Sudan. Now it was time to answer the questions: Is there a basin? Where is it? How large is it? How deep is it? The tool to answer those questions was the airborne magnetometer.

When we received word in mid-September that the interior licenses would be awarded to Chevron, we requested seven geophysical contractors to submit bids for aeromagnetic surveys in both the interior and Red Sea areas. A contract was awarded to Hunting Geophysical in November and, after preliminary aerial photography for navigation purposes, actual recording of airborne magnetometer data began on February 8, 1975.

The airborne magnetometer measures small variations in the earth's magnetic field. Igneous and metamorphic rocks, which form the "basement" beneath sedimentary basins, usually contain magnetic minerals that affect the magnetic field. Sediments, other than those derived from volcanic activity, do not. The measurements made with the magnetometer can be interpreted in terms of distance to the magnetic rocks or "depth to basement." The interval between the surface and the basement is assumed to contain sediments.

The first phase of the Hunting survey consisted of a series of flight lines called "triplets" (a triplet is three parallel lines approximately 1.5 kilometers apart) that were from 300 to 650 kilometers long and crossed the entire license area in a northeast–southwest direction. The triplets were flown at 40-kilometer intervals for the 1100-kilometer length of the area. As the recording of the first phase, totaling 60,000 kilometers of line, neared completion in April, Fred Flege and Paul Maton, a geophysicist from Hunting, met in Khartoum and made a preliminary interpretation of the data while the aircraft commenced flying the Red Sea survey. Their work indicated the presence of two basins! The larger, in the northwest, which we named the Muglad basin, covered an area roughly 250 by 700 kilometers and appeared to have more than 7000 meters (23,000 feet) of sediments. The second, named Melut, measured about 130 by 500 kilometers and had more than 4000 meters (13,000 feet) of sediments. Upon the aircraft's return from the Red Sea in May, another 21,000 kilometers of line were flown over the Muglad basin to improve the quality of the interpretation.

We were completely caught up in the excitement of exploration going right! An idea, which could best be called a geologically educated guess, was moving from dream to reality. If the airmag survey had indicated shallow basement under most of the area, the project would have been over. Now, almost certainly, we could look forward to a major exploratory program.

THE OIL MINING LEASE

In mid-March 1975, as the aeromagnetic survey was being carried out, Chapman returned to Khartoum to open negotiations to develop the mutually satisfactory Oil Mining Lease form promised by the director. From the time of our first contacts in the Sudan, all of the government negotiators had made it clear they wanted to avoid making any changes to the Petroleum Act of 1972. However, two of the requirements of the act (that the Oil Mining Lease holder be Sudanese, and that the government be issued share capital in the lease holding company) were particularly troublesome for Chevron since they could lead to loss of U.S. tax credits and deductions and result in the taxation by both the Sudan and U.S. governments of the same profits. The negotiations were further complicated by the fact that the new lease form would apply to the Red Sea permits as well as to the interior, so Texaco and Ball and Collins would also need to be involved in the negotiations and be satisfied with whatever was agreed upon.

By early September, a series of frustrating meetings had led to a set of draft agreements so complex as to be unworkable. We reviewed the situation when Chapman returned to San Francisco and we decided we should take a different approach. Rather than write an Oil Mining Lease on the basis of the 1972 Act, we would propose to the Sudanese that we enter into a Production Sharing Agreement of the type then in use in Indonesia and in Egypt, which was important given the close ties between Egypt and the Sudan. Under the terms of a Production Sharing Agreement, Chevron would serve as a contractor to the government for the exploration and production of oil and gas within the contract area. Chevron would furnish the necessary funds and expertise, and receive an agreed-upon share of the production as payment. One of the better aspects of the plan was that no significant change in the Petroleum Act was required; the agreement needed only a presidential order amending the Act to authorize the Minister of Industry to negotiate and enter into such agreements. We requested Tigani el Karib, our Sudanese legal counsel in Khartoum, to come to San Francisco in mid-September to assist us in drafting the proposed agreement.

Chapman returned to Khartoum in early October with the new proposal, which the Sudanese negotiators and our Red Sea partners quickly accepted. On October 12, 1975, a modern Production Sharing Agreement that was satisfactory to all the parties was signed in Khartoum.

ON WITH EXPLORATION

It was obvious that we had to make a decision about where to concentrate our efforts in the near term. Detailed evaluation of all of the prospective area would take several years and could not be justified until we had favorable answers to several questions about the nature of the sediments in the new basins—answers that could only come from drilling. On the basis of the most favorable terrain for field operations and ease of access, we decided to concentrate on the north half of the Muglad basin.

Although the airmag survey had found the basins, it was too dull a tool to do very much else. We needed a seismic survey to help us choose well locations, and given the high cost of seismic work, the short time available, and the size of the area, we had to have some means of focusing our efforts. The answer was a helicopter gravity survey.

Whereas the airborne magnetometer measures small variations in the earth's magnetic field, the gravity meter measures small variations in the force of gravity. These variations can often be related to the density of the rocks near the meter, and in turn the rock density can be related to the structure of the rocks. The gravity meter usually gives much more precise information than the magnetometer but is also much more expensive to use. In order to make gravity measurements at the precision needed, the meter must be standing absolutely still, in contrast to the airborne magnetometer, which records continuously as the aircraft crosses the prospect area.

In mid-September, 1975, a crew operating out of a tent camp at Aweil, on the west edge of the contract area, began recording a series of northeast–southwest gravity profiles across the northwestern portion of the Muglad basin. The lines were 16 kilometers apart, and had recording stations every 3 kilometers. The gravity meter was carried by a helicopter that landed at each station to make the measurement

Bill Chapman visiting with President Nimeiri at the conclusion of the negotiation of the Production Sharing Agreement, September 1975.

and then proceeded to the next station on the line. Approximately 3000 stations were recorded before February, 1976, when the work was halted by water-covered terrain on the south.

The first seismic operations in the interior began with a Petty Ray crew in January, 1976, near the western edge of the contract area southwest of the town of Babanusa. Operations were initially along Sudan's only railroad, a narrow-gauge track built originally by the British, which extended inland nearly 2000 kilometers from Port Sudan on the Red Sea coast across the northern portion of the contract area. The railroad allowed relatively easy access for the crew and simplified mobilization of equipment by rail. This part of the area was flat, sandy desert with scattered scrub trees and an occasional large baobab tree. It proved to be the easiest area of operations we encountered.

The original seismic base camp was at Babanusa, but it was soon moved 30 kilometers south to El Muglad. The camps were managed by a recently hired expatriate—Gordon Banfield, a New Zealander. Gordon was the first of a series of expats hired by Chevron to run interior bases and manage the logistical operation. These men were of various nationalities, widely varying ages, and unique personalities, but they had two traits in common. They all had African bush experience and they had the ability to stay in the field for long periods of time with minimum supervision.

Chevron's contract area ranged from dry desert in the north, to the Sudd, one of the world's largest Papyrus swamps, located 600 kilometers to the south. Between these two extremes were open grasslands interspersed with forests of thorn trees and black gumbo soil. The southern two-thirds of the area flooded during the rainy season. The rains generally started in May and much of the contract area, literally thousands of square kilometers, was covered with water for the next three to five months.

The area was also marked by distinct tribal boundaries. The most important was near the Bahr El Arab, a tributary of the Nile. To the north were the Baggara, a nomadic Arab tribe that moved their herds of cattle, goats, and camels north and south with the seasons, following the grass. With Chevron's arrival, the water wells drilled to support the oil exploration effort became magnets for these nomads and we had thousands of animals in the area of our rigs at times. To the south of the Bahr el Arab were the Dinka and the closely related Nuer, Nilotic African tribes. As a result of grazing disputes, the Baggara and the Dinka had been antagonists for centuries. Early in the seismic operations, Petty Ray recorded a long seismic line that crossed this tribal boundary, which was unknown to either Chevron or Petty Ray personnel. That afternoon the Baggara laborers laid out the geophones

Natives of El Muglad watch the departure of a Chevron Twin Otter aircraft.

for recording, but in the evening they quietly disappeared and returned to their villages. The next day Petty Ray had to began the job of hiring Dinka laborers to finish the line.

The initial seismic results were very encouraging, and in July 1976 the addition of a second seismic crew was recommended. The recommendation read in part:

> A helicopter-borne gravity survey...[has]...indicated numerous anomalies interpreted to represent major structural features and trends. Long [seismic] reconnaissance lines were initially programmed to give us insight into the nature of the basin, the type of structure present, and the reliability of the gravity interpretation...six hundred miles of reconnaissance seismic data have now been recorded. These data show very good...correlation with the gravity work. Stratigraphic inferences drawn from seismic velocity analyses are also positive. One single, broad, highly prospective gravity trend crossed by seismic data is over 300 miles in length...it is critical that we develop a...number of suitable drilling locations before bringing in a rig for the drilling program, which must be started in late 1977....In short, a multi-well pro-

gram will be required and we must develop data to ensure that the holes are located with sufficient geologic and geographic diversity to properly assess the basin's potential.

While the initial surveys were being carried out in the interior, Chevron's Khartoum staff had been engaged primarily in managing the joint Chevron–Texaco exploratory program in the Red Sea. By early 1976, two noncommercial gas discoveries had been drilled in the offshore, but hopes for something better were dim. With the decision to expand onshore seismic operations in the interior, there was a changing of the guard in Chevron's Sudan personnel. Jim Payne replaced John Sutherland as General Manager, and soon after that Dick Beam arrived as Manager of Seismic Operations. Later that year, Chester Arinder came to Khartoum as Drilling Manager. These men, along with José Echiverra–Government Relations, Barry Hughes–Aviation, Bob Nielson–Engineering, and John Kurylak–Accounting, made up Chevron's management team in Khartoum as the interior block became the primary operational concern. The remainder of Chevron's administrative support staff in Khartoum were primarily Sudanese who proved to be extremely capable in their jobs.

During that time, several philosophical decisions were made that served Chevron well throughout the early exploration of this very remote area. The logistics of providing supplies and equipment, for the most part, would be done using the existing Sudanese infrastructure—the narrow-gauge railroad and souk trucks with local contractors. Primary air support of personnel would be by company aircraft with company engineers for maintenance; and Chevron expatriates in Khartoum would live in the general community with no special perks or company commissary.

The initial seismic work by Petty Ray in early 1976 made it clear that some of the most prospective trends were in the southern areas, which were wet several months of the year. It would be critical that we obtain the data necessary to prepare for drilling in the south at an early date. The only problem was that neither the contractors nor Chevron knew much about the southern Sudan or how to work there.

In the early fall of 1976, Avo Mogossian agreed to take Jim Payne and Tony Rechner, a Chevron seismic supervisor, into the interior. Avo was, among other things, a big game hunter who conducted hunting trips into the interior of Sudan and he was experienced in the area. In return for air support and logistical costs, Avo acted as a guide for Payne and Rechner while they spent seven days in the interior determining how best to operate.

During that trip, Chevron made an ally who was to prove invaluable

Beyond Khartoum 179

Baggara tribesmen waiting to water their camels at Chevron drill site.

Dinka tribesmen observing helicopter-supported seismic crew in the marshes at the northern edge of the Sudd.

over the next several years. Beshir Abu Sinenna was an Arab trader located in the Nuer village of Bentiu. Beshir was a wealthy and highly respected trader who split his time between Omdurman and Bentiu and provided critical logistical support during those early years. One of the most significant operating decisions to come from Payne and Rechner's trip was the selection of Bentiu as our southern base camp.

Bentiu was located on a tributary of the Nile and was accessible mainly by boat. As a result, it was relatively isolated, as was much of the southern part of the contract area. An example of that isolation occurred during our first aircraft trip there. Our pilots emptied water from their ice chests on a small dirt air strip on high ground near Bentiu, because all the ice had melted. Splashing and laughing with great merriment, a group of Nuer children crowded around the chests as they drained. The children had never felt cold water before!

The second seismic crew, which was contracted from United Geophysical, began work in January 1977. The crew was equipped with special vehicles for operating under wet conditions and it was decided to have the crew work in the south through the 1977 wet season. While the crew could operate with helicopter support, the heavy material, such as fuel and dynamite, had to be stockpiled in supply dumps prior to the rains. To do this, Chevron hired two expats with African experience and seconded several longtime Chevron employees to work with these men. One of the Chevron employees was Jim Myers, who arrived in Khartoum in March 1977. Jim had spent most of his life in Denver, Colorado, and this was his first trip overseas. Within a few days of his arrival, he went into the field to begin supervising some of the stockpiling operations. Almost immediately, he had to arrange the evacuation by air of a native who had been mauled by a lion at his first dump site. Several days later, Jim was caught in the crossfire of a fight between the Baggara and the Dinkas. Myers was in the middle of his supply dump, which comprised dynamite and fuel in rubber bladders, while the two parties exchanged gunfire. Ultimately, all the dumps were completed on schedule and Jim returned safely to Denver with a better appreciation of African history and culture.

The 1977 wet season seismic program proved to be far more difficult than had been anticipated, resulting in low production rates and soaring costs. The necessary data were obtained to lay out the next dry season seismic program and begin preliminary planning for the drilling operation. However, the experience convinced us that in the future both seismic crews should work in the south during the dry season and in the north during the wet season.

By mid-1977, both El Muglad and Bentiu had become expanded bases of operations in preparation for the upcoming drilling program. Benji, an Indonesian national, was hired as manager of Bentiu. Over the next several years, Benji turned the base camp at Bentiu into a garden spot in central Africa. He had acres of gardens and flowers and routinely provided the best meals in Chevron's Sudan operations. And the drilling engineers could always find their missing drill pipe in the Bentiu fences and structures!

DRILLING BEGINS

In early 1977, a Parker drilling rig that Chevron had used for drilling programs in Madagascar and Kenya was brought to Port Sudan and was started on the 1200-kilometer journey by rail to El Muglad. The rig had to undergo significant repairs, but in October 1977, our first well, Baraka #1, was spudded 60 kilometers south of El Muglad using newly trained crews from the local villages. The well tested a large, highly faulted structure and encountered basement rocks at 2435 meters (7986 feet). No oil shows were encountered, but the well was encouraging because it penetrated more than 120 meters of organic shale that could generate oil if it were buried more deeply at another location.

The location for the next well, Unity #1, was 230 kilometers to the

Dinka laborers carry a replacement drilling line 80 kilometers to the wet season location of Baang #1 exploratory well. Heavy rains prevented trucks from reaching the location. July 1978.

southeast of Baraka and about 40 kilometers north of our camp at Bentiu. By that time, Chevron had contracted with a small construction company in London, called The Middle East Company, to provide construction and road-building support. Even with a good road, this move took 40 days using both Kenworth and the local souk trucks. Most of the delay was the result of inexperienced drilling crews, but we also learned to rely more on the smaller souk trucks with their local drivers. Within a year, our rig moves were reduced to 7 to 10 days, depending on the distance.

Amid hopes that a new era had come, the name for Unity #1 was chosen because of its location near the Arab–Black Africa boundary that had historically divided the northern and southern Sudan. The well was drilled on a major structure which, on the basis of seismic lines spaced at 10-kilometer intervals, appeared to have closure of about 260 square kilometers (100 square miles). The well was drilled to a depth of 4415 meters (14,483 feet) and encountered numerous minor shows of oil from 2280 meters (7500 feet) to total depth. A drill-stem test of a low permeability sand at 3430 meters (11,200 feet) recovered eight barrels of high-gravity, waxy crude oil. The business and diplomatic communities in Khartoum were full of rumors of a major discovery, and reports of a "lake of oil" hit the world press. Notwithstanding the rumors, Unity #1 was a dry hole and was abandoned in late June 1978, as the rainy season began.

While Unity #1 was drilling, a second Parker rig had been brought into the Sudan interior. About this time, Chester Arinder was transferred and was replaced by Oliver Brown as manager of drilling. With the oil shows at Unity, we felt it was critical to continue the evaluation effort in the south and decided to attempt to drill through the wet season with both rigs. The rig that had drilled Unity #1 was moved southeast to the location of Baang #1 ("Baang" is a Dinka word for "good luck") some 60 kilometers south of Bentiu in an area of perennial marsh at the north edge of the Sudd. The location was stockpiled, an elevated pad was built, and a successful rig move from Unity #1 to Baang was accomplished before the rains. However, while the rig was being erected and after the rains had begun, it was discovered that the drill line was defective. The replacement drill line was too heavy for our helicopters, fixed-wing aircraft could not land, and trucks could not move in the mud. The problem was solved by hiring 300 Dinkas who transported the 900 feet of steel drilling cable, by foot, on their shoulders like a giant snake. It took them three days with helicopter support for food and medicine, but the drill line arrived on location in time.

Baang #1 didn't live up to its "good luck" name; it was a dry hole with no shows, as were two other deep and expensive tests in the south. By the spring of 1979, both rigs were back in the northern part of the contract area drilling locations in the area west of El Muglad. Tension was building within Chevron Overseas and Standard of California. By now, Chevron's Sudan operations had expanded to two drilling rigs, two seismic crews, a gravity crew, a large construction group and an air force that consisted of two company Twin Otters, four helicopters, and six contract DC-3s. The six DC-3s were required in order to keep two flying at any given time. The original $15,000,000 expenditure commitment made to government for the first four years had become an actual expenditure of nearly $90,000,000. The question of finding a partner, which had been tacitly deferred following Texaco's refusal to join in the interior block, became a subject for active discussion. Chevron Overseas's reluctance to share the rare opportunity to control what might become a major new oil-producing province had to be balanced against just how much money Standard Oil Company of California wanted to place at risk in a single project.

The debate was again postponed when, in March 1979, oil shows were encountered in the Abu Gabra #1 well located 50 kilometers west of El Muglad. After being drilled to a depth of 4190 meters (13,744 feet), the well was completed in July, flowing 575 barrels of oil per day

President Nimeiri jumping over sacrificial bulls in a good luck ceremony at the Abu Gabra discovery celebration, June 1979.

from a thin, low-permeability sand at 2730 meters (8950 feet). In August, the Tabaldi #1 well, 20 kilometers north of Abu Gabra, encountered numerous shows and flowed oil at the rate of 135 barrels per day on a drill-stem test before being abandoned.

While the commercial significance of these results was certainly in doubt, they gave a much-needed boost to the Sudan operation and resulted in a wave of optimism by government officials and within Chevron. The Sudanese government hosted a major celebration at Abu Gabra. President Nimeiri, Minister of Energy Sharif El Tuhami, and Jim Payne of Chevron attended a rally of more than 10,000 persons. Speeches, the ceremonial sacrifice of cattle, and a large community barbecue were included in the festivities.

The event underlined the importance the government placed on the oil exploration program. The early efforts of Chapman, John Sutherland, and later Jim Payne and José Echiverra to develop an atmosphere of mutual trust and understanding between Chevron and the government leaders in Khartoum, Juba, and the villages of the interior had been very successful. We couldn't have hoped for a more supportive host government.

Jim Payne, with interpreter, addressing the crowds at the Abu Gabra celebration, June 1979.

ANOTHER LOOK AT UNITY

Following the abandonment in 1978 of Unity #1, we had in-filled the seismic grid on the structure from 10-kilometer line spacing to approximately 2.5-kilometer spacing. The new seismic data indicated that the crest of the structure was located approximately 13 kilometers south of the #1 well, with more than 100 meters of north dip from the structural crest to the #1 dry hole. The shallowest oil shows in the well had been encountered in a very porous sand at 2286 meters (7500 feet), but only water had been recovered on a drill-stem test of that interval. Below that depth, porosity decreased rapidly and tests of numerous shows recovered little except the 8 barrels of oil at 3430 meters (11,200 feet) that had led to the press reports of a "lake of oil" in the interior Sudan.

In early December 1979, with the dry season at hand, Chevron Overseas requested a meeting with the Standard of California exploration management to review a recommendation that one of the rigs be moved south for another test of the Unity structure. Such reviews usually were routine, generally supportive, and uneventful. Howev-

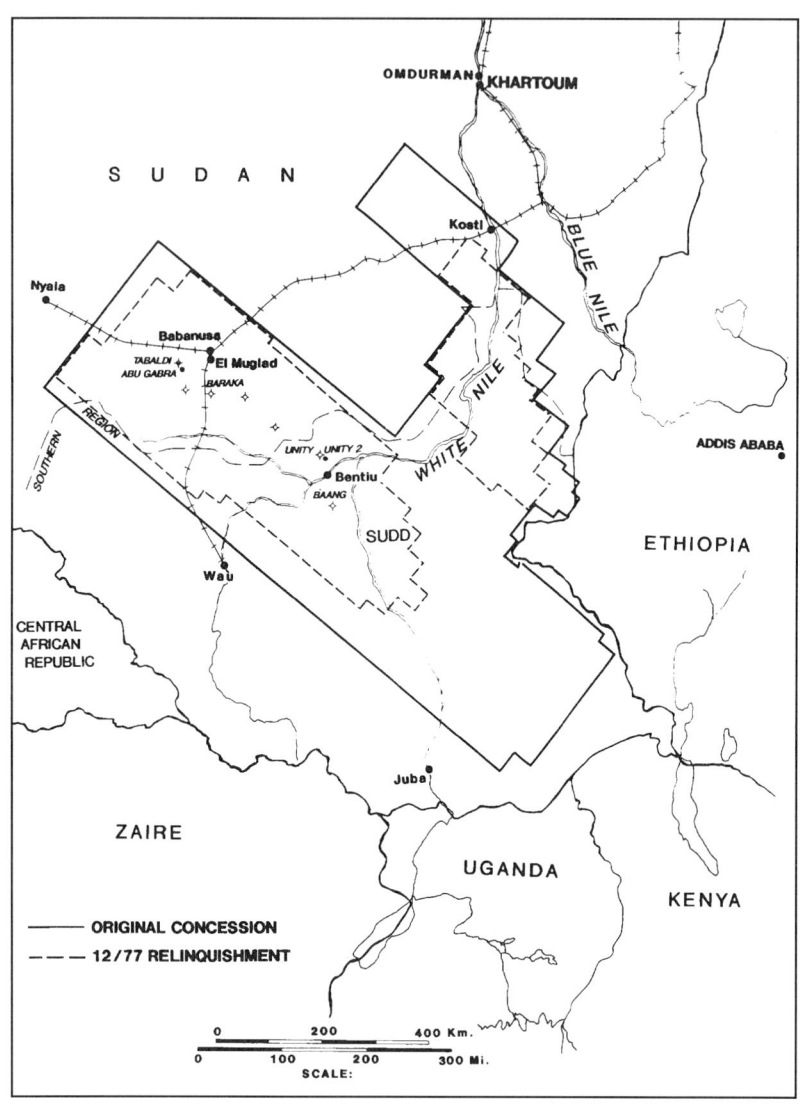

Map of the original December 1974 concession, showing areas covering the Muglad and Melut basins, retained at the time drilling began in late 1977. Note the exploratory well locations and boundary of the Southern Region.

er, on this occasion the mounting frustration with the Sudan program, especially that in the south, burst out. Almost as soon as the presentation began, the answer came back, "NO! Keep the rigs up north where the oil is! We're tired of deep, expensive, dry holes, tight sands, and rigs standing idle for months while we wait for the flooding to end!" It was nearly noon, and the participants scattered in some disarray for lunch.

Late that afternoon, we requested another audience to consider a compromise. The presentation concentrated on the 40 meters (130 feet) of sand at 2286 meters (7500 feet) in the #1 dry hole, which had oil shows, 20% porosity, and which might lie 100 meters higher at the new location. We agreed to forget about drilling again to the deep shows encountered in the #1. We would limit the depth to 2440 meters (8000 feet) and, if we were not successful, we would get the rig back to the north promptly.

DISCOVERY!

The drilling of Unity #2 began on the 29th of December, and in mid-January the drill bit began to encounter good oil shows at 1800 meters (5900 feet). The well was drilled to 2692 meters (8830 feet) and logged. Logs and cores indicated more than 70 meters of oil-saturated sandstone between 1800 meters and 2530 meters (8300 feet). With this encouragement, the decision was made to drill deeper but to test at least one zone before further drilling, in order to plan for the rapidly approaching wet season. Casing was run in the well, and on February 19, 1980 a test of the interval from 2393 to 2402 meters (7850 to 7880 feet) flowed oil at the rate of 2939 barrels per day!

When news of the test came into Jim Payne's office in Khartoum, two men were there who had been involved in the program from the time of Chapman's first visit—Yousif Sulieman, now Director General of the Geological and Mineral Resources Department, and Omar El Sheikh, now Chairman of GPC, the government oil company. Both had shared, with considerable patience and understanding, the many disappointments of the previous six years, and now they shared in the celebration. Within minutes, Yousif, Omar, and Jim were at the home of Sharif El Tuhami, the Minister of Energy, who immediately paid a visit to President Nimeiri to announce the good news.

The Unity #2 well was deepened to 3998 meters (13,112 feet) but found no deeper significant shows and was plugged back for further tests. Seven additional zones were tested between 1800 meters (5900 feet) and 2530 meters (8300 feet), and several flowed oil at rates of 1700 to 2600 barrels per day.

The Unity #2 discovery was the culmination of a 7-year journey. The journey had begun with John Miller's geologic observation and

Unity #2 discovery well during testing, April 1980.

the development of an idea, and had evolved into a complex field operation in a remote part of the world. The ultimate result was the discovery of several hundred million barrels of oil.

EPILOG

Although we couldn't know it at the time, the discovery of the Unity oil field occurred near the peaks of two unusual episodes in history. The first was the world energy panic, which caused oil prices to increase by a factor of ten during the seven years of exploration before the discovery at Unity. The second was the period of peace (well described in Chapman's first report as "peaceful, if not completely harmonious association") that existed between the people of northern and southern Sudan following the signing of the Addis Ababa Agreement in 1972. Both of these episodes were nearing an end.

The discoveries at Abu Gabra, and especially Unity, brought the understandable demand from government for an early oil project that would help ease the desperate economic conditions of the country. However, the field's remote location, the nature of the crude (which turned to wax at temperatures below 30°C [86°F] and was difficult to transport and refine), and the need to define sufficient reserves to justify the necessary investment, meant the project would take time.

Meeting with President Nimeiri after completion of the Unity #2 discovery, May 1980. From left: President Nimeiri; Sharif el Tuhami, Minister of Energy; Al Martini, Vice-President for Exploration, Chevron Overseas Petroleum; Jim Payne, Managing Director, Chevron Sudan; Swede Nelson, President, Chevron Overseas Petroleum.

The exploration program was expanded with another drilling rig and two more seismic crews, including one working in the Sudd. Tugs and barges were built to transport equipment and supplies on the Nile. By late 1983, construction of a pipeline to the Red Sea was about to begin. Line pipe manufactured at mills in Italy was on its way by ship to the Sudan.

In the early 1980s, however, the Nimeiri government's relations with the south had begun to deteriorate. This was accelerated in 1983 when presidential actions reduced the political power of the Southern Region by subdividing it, and imposed Islamic law on the non-Muslim south. While it was not the principal cause, the question of what oil project and where reopened the broader question of whether the south felt it would ever get a fair share of the fruits of economic development from any government in Khartoum. A new generation of "angry young men" rose in the south. Throughout 1983, various rebel groups were reported to be forming, reinforced on occasion by deserters from Sudanese Army units.

On the morning of February 4, 1984, rebel forces attacked the Chevron operations base near Bentiu, using automatic weapons and killing three employees of Chevron's contractors. The base was evacuated. The rebels also attacked the camp of a French contractor engaged in the construction of the Jonglei Canal, a major Nile River development project, 170 kilometers to the east. Their purpose was, clearly, to challenge the authority of the central government and deny it any benefit these projects might bring, whatever the cost to North or South.

In the years that have followed, several new governments have held power in Khartoum and the rebels remain in the south. During the same period, the price the world will pay for energy has dropped by half. The story of producing the petroleum found in the interior of the Sudan remains in the future.

[**Note**: A complete technical description of the interior Sudan program and geology may be found in:

Schull, Thomas J., 1988, Rift Basins of Interior Sudan: Petroleum Exploration and Discovery: *AAPG Bulletin*, v. 72, #10, p.1128-1142.]

Pakistan

First Oil in the Sind

By Herb Young

INTRODUCTION

Although the discovery of oil in the province of Sind in May 1981 may not have been a significant event on a worldwide scale, it was very important in the context of oil exploration in Pakistan. Not only was this discovery at Khaskeli the first oil discovered in the Sind, it was also the first oil discovered in Pakistan outside the Potwar basin in the northern part of the country. The Potwar had been producing since 1915 and its fields had long been in decline. Exploration throughout Pakistan had stagnated. However, the Khaskeli discovery changed this—it proved oil in an area that at the time was not generally considered attractive for exploration, and in so doing drew the attention of the international exploration community to this previously neglected area. In large part, the success in the Sind was responsible for the upsurge in exploration activity in Pakistan in the latter part of the 1980s and into the 1990s. Perhaps from this renewed interest will come the world-class discoveries that we hoped for when the Khaskeli-1 was drilled.

For me, the Sind discovery had been a long time in coming. More than 16 years had passed between my first visit to Pakistan and the day I stood on the rig floor while the drill was cutting the first commercial oil pay sand in the Sind. I believe the emotional links I had formed with Pakistan during my early visits encouraged me to persist in looking for oil there, and that the anticipation of success had been a powerful motivating force and a spur to creativity. I was particularly gratified when, in March 1982, the Pakistan government decorated me with the Star of Pakistan. The citation read, "In view of Mr. Young's dedication and perseverance in the field of Oil Exploration in Pakistan, the award of SITARA-I-QUAID-I-AZAM has been conferred on him."

THE SIND

The Sind is the southernmost of the four provinces of Pakistan, lying along the Indus River between the mountainous province of Baluchistan in the west and the deserts and salt flats of India to the east. In early history, Sind was the name of the river rather than the region. When Alexander conquered this flood plain in the fourth century B.C., he referred to the river as the Sinthos and applied variations of this word to the region and its people. Subsequent transliterations into Latin, Arabic, and later, into European languages, provided the names Indus, India, and Hindu. Today, the term Sind designates the province, while Sindhi denotes the inhabitants and their language.

In the present day, aside from the industry in the cities of Karachi and Hyderabad, the Sind is supported by agriculture of varying intensity. The climate ranges from semi-arid to arid, but most of the Lower Indus alluvial plain is irrigated and arable. The principal drawback to fertility is salination, except on the eastern margin of the province where blowing sands have formed a desert.

Farming has a reasonably modern appearance in the fertile and heavily populated areas close to the Indus River. In the more remote areas, life is quite rustic and the farming implements of a timeless past are common. Fields are often cultivated with wooden plows pulled by bullocks or water buffalo. Water wheels being turned by camels can be seen lifting water to irrigate the fields.

Throughout the remote areas, there are collections of huts made from reed mats and thatch. These communities are entirely enclosed within a wall of thornbushes and probably house kin groups. I have been told that those houses having conical roofs belong to Hindus who remained on the West Pakistan or Moslem side of the international border after Pakistan's partition from India in 1947.

The Lower Indus Valley in the Sind was one of the cradles of early civilization, as the ruins of Moenjadaro prove. This is a well-known archeological site and attracts many visitors, but throughout the entire flood plain of the Indus there are many vestiges of previous cultures, little visited and unmarked on tourist maps.

Geologically, the Sind Province lies within the Lower Indus basin, a cratonic marginal basin flanking the northwest side of the Indian shield. In the Sind, the basin consists principally of a gently westward-sloping shelf where the section is dominated by a great thickness of Mesozoic sediments unconformably overlain by volcanics and a relatively thin section of Tertiary sediments. To the west, in the fold belt of Baluchistan, these Tertiary sediments attain considerable thickness. Underlying the Mesozoic in other parts of Pakistan are Paleozoic sediments, and that is probably also the case in the Sind, although no section of Paleozoic age has been drilled or crops out in the province.

Petroleum exploration of the Lower Indus basin in the Sind was very sporadic between 1893, when the first well was drilled, and

Herb Young being decorated by President Zia Ul Haq.

World War II. After the war, however, exploration picked up. Exploration of the Lower Indus was perhaps encouraged by the perceived similarity of this basin to the prolific basins of the foreland shelf of Arabia, or by the fact that the Indus basin was a link in a great chain of petroliferous basins of the Tethyan seaway stretching from Europe and North Africa, through the Middle East and connecting with those producing basins of Southeast Asia and the Sunda Archipelago.

The postwar exploration surge in the Sind drew to a close with no oil having been found, but with considerable gas reserves being established at Sui and Mari. This set the stage for the next cycle of exploration, in which I was involved personally.

EARLY VISITS TO PAKISTAN

My first experience in the Sind was a visit to Karachi in 1965 as a geologist for Sun Oil Company. Sun was in the middle of a three-well drilling program that was the first offshore drilling in Pakistan. Although unsuccessful, it was a well-conceived attempt to test the oil potential of one of the world's great deltas. I remember the visit vividly because it was also my first experience with Ramadan, the Moslem

Hindu community near Khaskeli in the Sind.

month of fasting.

When I arrived in Karachi, I was interested in seeing the sedimentary section that the offshore wells would encounter. Two of the Pakistani geologists working with Sun at that time, Iqbal Kadri and Khaliq Qureshi, volunteered to take me into the hills to see the outcrops. I didn't realize we would be fasting and it was somewhat of a shock to my system to climb about all day in the Pab range in the hot sun without having anything to eat or drink. Despite my torment it was the beginning of a long-term friendship with these two geologists, and we later crossed paths in a number of different places in the world.

In 1975, I left Sun and joined Union Texas Petroleum as Exploration Manager in Singapore. One of the first things I did in joining the company was to write a memo urging consideration of exploration of the Lower Indus basin in the Sind Province, Pakistan. Bob Stover, our General Manager in Singapore, immediately became interested. Whether his interest was for geologic reasons or the thrill of working in a new country I cannot say, but at any rate, we planned a trip to Islamabad to speak to the Director of Petroleum Concessions and to review what data were available in the government files so that we could further evaluate the basin.

In that first visit to Islamabad, Bob and I were able to put together a reasonable data base documenting the results of the previous exploration in Pakistan, particularly the Lower Indus basin. On this trip, not only did we find information in the government files, we also found it useful to visit with Pakistani geologists and geophysicists. They were eager to discuss in general terms the regional geology and exploration problems in Pakistan.

Islamabad was built in the late 1960s as a new seat for the capital of Pakistan because General Ayub Khan desired to move the government from Karachi in order to distance it from commercial interests. He chose to build it in the northern part of the country near Rawalpindi, on a site framed by the Margalla hills, which rise eastward to meet the mountains of Kashmir. The setting is lovely and the area usually has such pleasant weather that it's a joy to visit. But perhaps more impressive than the scenery at Islamabad are the historical sites of Northern Pakistan, which seem to have that mystical aura found on the sub-continent. Two sites that must be visited are the Khyber Pass and the ancient city of Peshawar, which guards the entrance to the pass.

This was Bob's first trip to Pakistan, so I felt I should introduce him to the roots of the country. I wanted to show him the relics of a culture that pre-dated Mohammed, Christ, and Buddha, and whose early history involved the Hindu legends of the Mahabharata. We hired a car and driver, and by way of Taxila, Attock, and Nowshera, took a day trip to Peshawar and the Khyber Pass.

The road we took is, in part, the route of the Grand Trunk Road, which Afghani and Mogul rulers of Northern India built to link Kabul and Delhi. Not far beyond Rawalpindi, this road crosses a ridge on whose crest is a tall, slim monument erected in memory of General John Nicholson, who was killed at the age of 36 while assaulting the gates of Delhi during the mutiny of 1857. The monument is one of the many reminders, in this part of the world, of the days of the British Empire. A little farther along the road one finds part of the original Grand Trunk Road, which is well preserved because it was paved with the tough nummulites limestone of the Eocene Margalla Hills formation, which crops out nearby. That section of the Grand Trunk Road was described by Mountstuart Elphinstone in his account of his negotiations with the Afghanis in 1809.

Elphinstone had been sent to Peshawar not only to negotiate a treaty, but also to find out as much as he could about the Afghani's way of life. The British until then had little contact with the tribes beyond the Indus and lacked information of this strategic area between the Indian sub-continent and what they perceived as the Russ-

In 1809, Elphinstone described this remnant of the Grand Trunk Road near Rawalpindi.

ian menace to the North. To make a map of the lands controlled by the rulers in Kabul was a necessity; to do it without sending surveyors into the region was a problem. Elphinstone interviewed countless caravaners, itinerant merchants, nomads, and whatever other traveller he could find. Through these interviews he determined how many days of riding it took to get from one village to another, in which direction the rivers and mountains were, and so on. His product was a detailed work of outstanding accuracy.

Continuing our trip to Peshawar, we passed the ruins of Taxila, the ancient capital and seat of learning of the Gandhari tribes who were flourishing a thousand years before Christ and who were still doing well at the time that Alexander the Great passed through, in 327 B.C. From there we headed to Attock, crossing the Indus River where it begins its plunge through the gorge it has incised in the Potwar plateau.

Peshawar itself stands on the edge of the plains, close enough to the hills to protect the entrance to the Khyber Pass. This has always been an important city stronghold, located as it is in the zone of conflict between those who ruled the plains of India and those who came to conquer from the mountainous regions to the northwest. For the

At the entrance to the Khyber Pass, Charles Moerbe looks back towards Peshawar.

British, it was the base for their forays and expeditions into the Khyber Pass and Afghanistan.

The most interesting part of Peshawar is the old city with the Qissa Khawani, the bazaar of the storytellers. The streets here are crowded with colorful Pathani tribesmen, their dark beards framing grey eyes, features that are perhaps inherited from previous Greek invaders. Their robes and turbans have looked the same for centuries.

After Peshawar, we soon left the plains and began the long, slow, winding ascent up the foothills into the Khyber Pass. All along the road are plaques and small monuments to the British military units who saw action here. The tensive atmosphere in the pass is frightening—when you look up into the slopes above you, you can easily imagine snipers behind the boulders and you can almost hear the deafening, reverberating sounds of past battles. It is said that during the thousands of years that this pass has been the object of battles, every stone has been drenched in blood, and I believe it. The thrill of going through this pass has little to do with a vista of natural beauty, for the hills are dry and bleak and largely devoid of vegetation. Rather, it's the romance of being at a focal point in the history of a great and mysterious land. The starkness is daunting. The collective apprehension of all those who have passed through these clefts still hangs in the air. You are acutely aware that if you leave the government-protected area of the road, you have gone beyond the pale and your life could be forfeit.

On the way back to Islamabad, I thought a lot about those who had been this way before—of Alexander the Great, of Barbar and his Mongol hordes, of the diplomatic agent Elphinstone, and of course, of Kipling. I was caught up in the feeling of this land and I became more determined to make my own little mark on its history. I had begun this trip to give Bob an insight into the culture of Pakistan. I accomplished that, but in so doing, I became even more addicted to its romance myself.

EXPLORATION CONCEPT

Exploration concepts in the Sind and the adjacent provinces of Pakistan have followed the same history of development that exploration has followed in other parts of the world. During a period of almost a hundred years, ideas progressed from drilling on oil seeps to defining anticlinal structures from surface geology, through the use of the torsion balance and finally to the application of seismic data interpretation.

In 1866 and again in 1885, wells were drilled on seeps in the Punjab and Northwest Frontier Province. In 1893, the first well in the Sind, at Khaipur, was sited on an Eocene outcrop protruding from the Indus alluvial plain; the outcrop was evidently the surface expression of an anticline. (Or perhaps the well was placed at that spot because it was conveniently within the confines of a railroad repair yard.) In 1925, Burmah Oil drilled a second well in the Khaipur area, this time on a gravity feature defined by torsion balance. It and the preceding well at Khaipur were dry. Finally, in 1956, the structure was again tested when Burmah Oil drilled a seismic prospect. Gas was discovered, although not in commercial quantities.

Burmah Oil's drilling at Khaipur in 1956 was not an isolated one-shot event, which much of the previous exploration had been. Rather, it was part of a larger exploration effort that involved a number of companies and that had been encouraged by the government of Pakistan beginning shortly after partition from India. This activity, lasting from the early 1950s to the mid-1960s, was the first intense exploration of the Lower Indus basin. Besides Burmah, it involved such companies as Stanvac, Sun, Tidewater, and Hunt. The Stanvac area, which covered the southern part of the alluvial plain, almost as far south as the man-

Location map showing the Lower Indus basin, the Potwar basin, the Sind, and the Badin Block.

grove swamps of the present day Indus Delta, later became the Badin Block of Union Texas in the subsequent cycle of exploration.

The geologic information that was gained from the period of exploration starting in the 1950s formed the basis for the play I recommended to Union Texas. Those were the data that Bob Stover and I had gone to Islamabad to get.

When we returned to Singapore, we had the data to support a proposal suggesting that our home office management in Houston apply for an exploration license in the Lower Indus basin. Based on what we then knew of the basin, we were able to put forward a geologic model that showed possibilities of finding large oil accumulations. That basin, we thought, deserved a more thorough exploration program with newer geophysical techniques, and that is what we proposed to management.

For an area about which little is known of the geology, developing a geologic model or an exploration play concept is very much like trying to put together a jigsaw puzzle from which most of the pieces are missing, and lacking a picture on the box cover by which to go. The art of being successful in the game of exploration is being able to visualize the geologic scene with the least number of pieces, or data points, and then being able to convince others that you have a plausible picture.

Initially, I had narrowed the area of interest in Pakistan to the southern part of the Lower Indus basin between the border of India and the Indus River, because the sedimentary section encountered in five wells drilled by Stanvac Pakistan surely indicated that a major depocenter had persisted in just that zone throughout the Cretaceous period. Shows of oil in the Lower Cretaceous Sembar formation indicated the presence of source rock. Great thicknesses of potential reservoir sand, perhaps deltaic in origin, were found sandwiched between Lower and Upper Cretaceous shales and were roughly equivalent in age and character to the Burgan and Zubair sands in Kuwait. I reasoned that if this recipe for oil generation had worked in Kuwait, why not in Pakistan?

Of some concern, of course, was that these five Stanvac wells were dry. The answer had to be that the wells were not valid tests of good structures. The offhand dismissal of previous exploration reasoning and interpretation often is overplayed by explorationists trying to promote their own ventures and can be dangerous if unsupported by a rational explanation. In order to gain approval of the Pakistan proposal from Union Texas management, I realized I would have to show convincing arguments that these wells did not, in fact, condemn the prospect.

Our area of interest was under the flat alluvial plain of the Indus where there were no rock outcrops to give a clue about structure, so it was necessary to rely completely on geophysics. I knew from my association with the Sun exploration program that this was a problem area for seismic, primarily because of the basaltic flows of the Deccan Trap, which overlie the Tertiary–Cretaceous unconformity surface. My premise was that with the seismic technology of the early 1950s, it was not possible to get mappable reflections from the target Mesozoic section below the highly reflective Deccan Trap. Hence, the Stanvac wells hadn't tested the true oil potential of the area.

To demonstrate this, I drew a cross section of the first four Stanvac wells, which had been drilled in almost a straight line. The section showed a relatively flat Tertiary section overlying an erosional unconformity. Below that was a beveled Cretaceous section, which became more profoundly eroded towards the east, or updip, direction. On that section I also plotted a Cretaceous horizon that had been mapped by Stanvac, which I obtained from the government files. The Cretaceous horizon from a map by Stanvac, using seismic interpretation made prior to the drilling of the wells, in no way corresponded to the well data.

The map showed little relief, whereas the wells demonstrated considerable regional dip with vast sections of Cretaceous sediments missing by erosion, from one well to the other. It was apparent from this cross section that the reflection mapped as Cretaceous by Stanvac was almost certainly the first multiple of the Deccan trap reflection. It conformed to the flat-lying Tertiary structure and had nothing at all to do with the Cretaceous structure.

Establishing that the previous wells had not been valid tests did not mean that there were, in fact, closed structures present. However, we had another clue besides the questionable seismic—we had the gravity map. It showed a belt of northwest–southeast-trending anomalies that conceivably could be the expression of Cretaceous structure. I felt we had sufficient reason to believe that the gravity anomalies were actually caused by Cretaceous structures, and that that assumption should make us comfortable enough to make the play.

The geologic map of the Kutch area of India, which is adjacent to the Sind, shows a number of large, faulted anticlines with northwest–southeast trends, the crests of which have Cretaceous and Jurassic sediment outcrops. When I first became aware of these structures, I realized that their trend lined up with the grain of the gravity anomalies on the Sind side of the border. If these anticlinal structures were exposed on the surface in India, I reasoned, why couldn't their extension, as indicated by gravity, be in the subsurface in Pakistan?

In simplest terms, frontier exploration consists of looking for areas where trends of favorable stratigraphy intersect structural trends, forming what is called a "Sweet Spot." The geologic model we had put together from bits and pieces of information showed a northeast–southwest zone of favorable Cretaceous sediments crossed by a northwest–southeast structural belt, a potential "Sweet Spot."

The concept was simple but testing it would take some time. The first step would be to get a concession on the acreage and then shoot a seismic program to see what useful data from below the Deccan Trap could be obtained. If structures were found, it would be up to the drill to determine if we had the rest of the geologic picture right.

MAKING THE DEAL

In the fall of 1975, we recommended to our Houston office that the company apply for an exploration license covering a block of 12,950 square kilometers (5000 square miles). This recommendation did not seem to cause a great deal of excitement and I imagine our management were somewhat at a loss as to how to deal with it. The Union Texas International exploration staff in Houston was quite small at that time, and none, as far as I knew, had any personal experience in Pakistan. John Kennedy, the company's Manager of Exploration for the Mediterranean area, had been with Amoco in Pakistan for some time. When John was asked his views concerning our chosen block of acreage in the Badin area, however, he voiced enthusiasm about the prospects of the country in general but had his own favorite areas that

The Secretariat, one of the first buildings to be constructed in the new capital of Islamabad. The Badin Concession Agreement was negotiated here.

unfortunately didn't include the Lower Sind. As a new employee with the company I had no base of credibility myself. What I needed at this point was an advocate in Houston to support the proposal.

At just that time, Jesse McCollum, the president of Union Texas, visited the Singapore office on his way to Indonesia, where the company had an interest in a large gas discovery then being developed. Jesse had been somewhat responsible for my joining Union Texas—I had known him prior to leaving Sun Oil and he was the person I had called when I was looking for a new company to join in order to stay overseas. During Jesse's visit to Singapore, I had an opportunity to discuss with him my concept of the geology of the Lower Indus basin. He evidently went away impressed with what I had to say. At any rate, things began to happen after his return to Houston. The proposal apparently had acquired an advocate, while I had probably estranged the entire Houston exploration staff for having worked around them!

An economic analysis of the prospect was run in Houston and it showed that the field would need a reserve of at least 100 million barrels in order to make a reasonable rate of return. Although I was not particularly concerned with what some of the staff might have considered excessively high expectations, I did manage to convince our engineers to use fewer wells, higher well rates, and lower pipeline costs in their model. That, of course, made the economics look a lot better.

A swinging bridge over the Swat River.

Signing the Badin Concession Agreement. Left to right: Shafiuddin, Young, Mashiuddin, Shehzad Sadiq, and Mashiuddin.

In May 1976, corporate approval was obtained to apply for the Badin block. Phil Brown, the Union Texas head landman, and I went to Islamabad to file the application. It was a very simple procedure of filling out a form and paying a fee, the equivalent of approximately $750.00. I paid this amount with cash out of my pocket so I could put it on my expense account, thereby in a sense placing my own stamp of ownership on the project.

When we filed the application with the Director of Petroleum Concessions, M. Shafiuddin, we had considerable discussion with him and his deputy, A.R. Memon, about the configuration of the block. Memon was anxious that we explore the Jurassic prospects as well as the Cretaceous, so he induced us to add additional acreage in the southeast to cover this objective also. This larger block covered 17,871 square kilometers (6900 square miles).

The next step in the process was to write a draft Concession Agreement and a Joint Operating Agreement to form the basis for negotiation. Pakistan had a model agreement, but since they had indicated that most terms were negotiable, we took them at their word and wrote a draft agreement from scratch. I actually had little to do with the writing of the documents other than to suggest what terms might be attainable. The draft was completed during the summer and submitted to the Pakistan government in October. We were told negotiations would be held in November.

In November we were called to Islamabad. Dan Spencer of the Houston Legal Department, Charles Moerbe from Land, and I made the trip to Pakistan to begin negotiations. Because Islamabad was then a city still under construction, the preferred accommodations were at the Hotel Inter-Continental in Rawalpindi. This hotel is on a beautiful tree-lined road called the Mall, which to the east heads to Lahore, and to the west to Peshawar. Supposedly, it is on the site of the old Grand Trunk Road. In the mornings, a very picturesque procession of horse carriages, or Tongas, passes by the hotel, mostly taking children to school. From reading Kipling, one can imagine that the scene must have been similar a hundred years ago.

The negotiations were conducted in the Secretariat building in Islamabad, which required about a half-hour's commute by car from Rawalpindi. In making this trip, I soon found that the Tongas I had admired were more picturesque when viewed from afar. My admiration swiftly disappeared when a group of them jammed together in front of me, blocking all traffic.

The meetings went rather well and I don't recall that we had any major crises. Both sides entered the give and take of the negotiation in good spirits and we even had a few laughs. For instance, the Union Texas lawyers in Houston, who had drafted the agreement, had been careful to specify that any instance in which an unsolvable conflict might arise should be referred to The Hague under the rules of arbitration of the International Chamber of Commerce. One of the annexures of the Concession Agreement was the Form Mining Lease to cover production operations in case of a discovery. The form of this lease was taken from the 1949 Petroleum Rules, which our lawyers had modified to suit the present situation. One of the clauses of this lease agreement provided that the concessionaire had the right to clear 10 acres of brush land to provide for a corral for draft animals, and the compensation for the land would be agreed between the company and the landowner. The Pakistani negotiators were quite astonished at Union Texas's proposal that if the company and the landowner could not agree on the compensation for this land, the matter would go to The Hague. We gracefully gave in to them on this point.

On one of the weekends during a lull in the negotiations, we had time to take a trip to the Swat Valley, one of the several Shangri-la like valleys in Northern Pakistan that run deep into the lower ranges of the Himalayas. Like the travellers in James Hilton's *Lost Horizons*, we also started our trip from Rawalpindi, but by car rather than by plane. To get to Swat, one must go through the Malakand Pass, where a young Army officer named Winston Churchill had once been stationed. The narrow road, which snakes up the mountainside and through the pass, is dominated by wildly careening, bizarrely decorated trucks and buses that try to bully all smaller vehicles off the road. What a relief it is to get through the pass to the peaceful, quiet valley of Swat on the other side!

When our negotiations in Islamabad were completed, we returned to Singapore and Houston to wait to be called back for the signing. We waited and waited. Not having heard from the DPC within the expected time, I became concerned that some hitch had occurred. Peet Stilley, our negotiator in Singapore, was dispatched to Islamabad to see what was going on. After a few days, Peet was able to get a phone connection to Singapore. He had found out that a young lawyer with the government was questioning whether oil was a mineral, because our request was to be included under a law that protected mineral development. Peet asked me if oil was a mineral, and if so, could I give him a reference. After a few moments of research, I found a dictionary that said, "Oil, usually considered a mineral, ..." That evidently was all it took to get the government's review of the agreement back on track. When Peet returned to Singapore, he described the incident as being "a little spice in the soup of life."

The signing was finally set for the latter part of April. There was an economy drive at that time to cut down international travel expenses, so it was decided that I should handle the event by myself. Unfortunately, a lot of last minute details had to be taken care of before signing and for awhile I thought I wasn't going to make it. A half-hour before signing, I was still in the customs shed at the airport trying to clear the company seal!

On April 22, 1976, the minister of petroleum, Shehzad Sadiq, signed the Badin Area Petroleum Concession Agreement on behalf of the president of Pakistan. I signed the agreement for Union Texas with the director of petroleum concessions, M. Shafiuddin, witnessing my signature. There was no one on my side of the table to do it.

Back at the Hotel Inter-Continental in Rawalpindi a few hours after the signing, I watched the news ticker giving the details of Union Texas's having just signed a new exploration agreement in the Sind. Also coming over the news was the story of PIA, the Pakistan airline, going on strike.

A general election had been held in Pakistan in March, which Prime Minister Zulfiqar Ali Bhutto's party had won easily. The opposition, however, would not accept the election and serious civil unrest resulted,

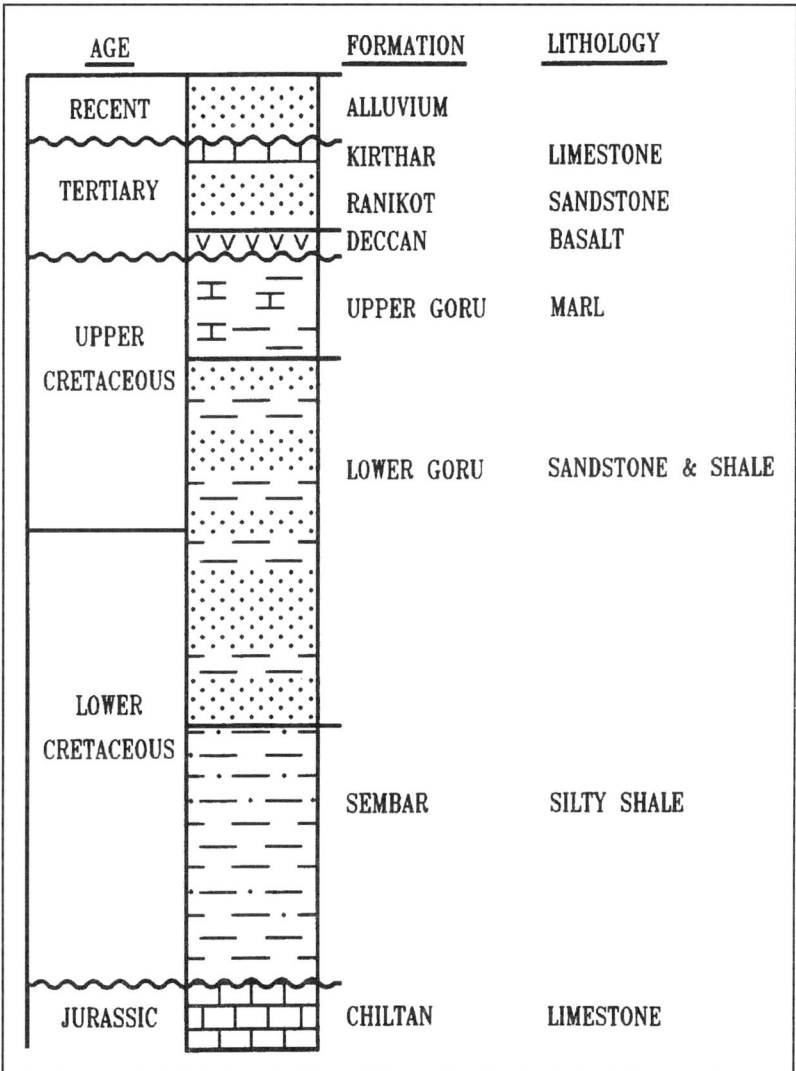

Badin area generalized stratigraphic column.

part of which was the PIA strike. One of the oil people, who was staying at the Inter-Continental Hotel, like me, wanted to return to Karachi and decided that he would go by train. I did not want to spend the time doing that, so I hired a car to drive me to the Indian border, intending to cross between Lahore and Armritsar. From there I expected to fly on to

New Delhi and Singapore. The driver took me to the border and left me to carry my heavy suitcase across a cane field to the Indian border check point. My car drove out of sight, returning to Rawalpindi, as I headed for the Indian immigration post. I approached a white-bearded Sikh immigration officer wearing a khaki turban, and presented my passport to him.

"Where is your visa?" he asked me.

"I don't need a visa," I replied. "I have been to India many times and have never needed a visa."

"That may be so when you come by air," he informed me, "but when you arrive by land you don't enter without a visa."

I dejectedly walked back to Pakistan through the waves of heat and the dust of the road, wondering how I was going to get to Lahore and from there to Singapore.

Luckily, I managed to find a car in a nearby village. It took me to Lahore, where, upon arriving, I checked into a hotel. I went to bed intending in the morning to try to get an Indian visa or a train ride to Karachi. When I awoke, I found that a 24-hour curfew had been imposed. I was going nowhere. The bad news was that the telex operator had been caught in the curfew at his home, so I couldn't inform my office or my wife where I was. The good news was that the tennis coach was at the hotel and couldn't leave. I managed to get in a lot of tennis while I waited.

After a few days, I had the good fortune to get a ticket for a special PIA flight to Karachi. A lot of people were trying to travel, and through the sheer good luck of being in the right place at the right time I managed not only to get a seat on the plane, but also to get a ride in a military vehicle to the airport. The airline was selling tickets for the plane but it wasn't providing transportation to the airport, and no civilian transport was allowed on the road. I learned later that my friend who had gone by train to Karachi had ended up walking the last 14 miles of the trip because a mob stoned and derailed the Khyber Mail.

I was quite relieved when, finally, I made it back to Singapore with the signed Concession Agreement.

EXPLORING THE SIND

The Concession Agreement we had made with the Pakistani government gave Union Texas a 90% interest in the Badin Block and the remaining 10% interest went to the Pakistani government, represented by their state oil company, OGDC. In addition, upon the first discovery, the government had the option to increase their interest to 40% by paying their proportional share of back costs. The 90% interest was more than Union Texas wished to carry in this exploration project. Therefore, my first job after signing the agreement was to find a partner.

I contacted Cities Service's regional exploration office in Singapore because I felt their exploration objectives resembled ours, and Les Beddoes, their Exploration Manager, was a good friend. Les invited me to lunch and I brought along a few work maps to show him what the play concept was about.

The Cricket Club where Les and I had lunch was similar in style to a great many of the British clubs built in the 1870s in Malaysia and India. These clubs all seemed to date to the same period of time, which probably corresponded to the opening of the Suez Canal in 1869. After that, a flood of British colonists had arrived in South Asia. Perhaps Queen Victoria only had one architect to send along to these outposts, and he was loath to experiment! All of his structures were wooden frame and painted black and white. A hundred years later, however, when Les and I were lunching, the clubs had acquired tremendous charm and we all felt it a pity when they started to be torn down and replaced by modern structures.

The playing fields in front of the Cricket Club had always been called by their Malaysian name, Padang. It was here, after the fall of Singapore to the Japanese in 1942, that the remnants of the British army had been gathered together to begin their march to Changi prison.

Les became enthusiastic about the Badin play and it wasn't very long before Cities Service agreed to take an interest in the concession equivalent to the interest held by Union Texas. Although I had shown the deal to several other companies in the meantime, none could act fast enough to make a decision before Cities agreed to the project.

The Concession Agreement called for us to spud a well before the end of the second contract year if we wished to continue into the second exploration period. The agreement's structure had an initial period of two years with a minimum financial obligation, followed by subsequent extensions, each of which had its own financial commitments. These commitments were high enough to ensure a continuous exploration program but could easily be exceeded if an aggressive exploration program were carried out. The short fuse on getting started on a well, however, meant that no time could be wasted in getting set up in Pakistan and getting started on the seismic program.

Ken Buss, our Administrative Manager in Singapore, was sent to Karachi to find office space and hire an office manager to get things rolling. Unfortunately, because of the civil disturbances the curfew was still in effect for most of the day and getting about town was difficult.

Ken did very well in making contacts with other operators, who helped out. Marathon Petroleum was particularly helpful because they had just finished an unsuccessful exploration of the Makran Coast and were about ready to pull out of Pakistan. In doing his job, Ken had to get special military passes to allow him to travel around Karachi during the curfew to look at office space.

Of the locations Ken recommended, we finally settled on a residence near the hotel area and not far from the main part of town. It was a large house with ample grounds that were taken up for the most part by well-established lawns and garden beds, and with coconut palms around the property perimeter. It provided a very pleasant working environment and served as our office for quite a few years before we outgrew it. One of the attractions was a fish pond, where our attempts to raise goldfish were thwarted by birds who enjoyed fishing. We joked that our first production in Pakistan was fifty coconuts.

When we rented the house, the owners asked us to take on some of their household staff, which we gladly did. Two of these people are still working for Union Texas 14 years later, while the others have since retired. Hassan, a driver we hired, thought he was unique among the drivers in Karachi because he had once chauffeured Chiang Kai-Shek when the general was on a visit to Pakistan.

After partition in 1947, the new State of West Pakistan had changed many of the street names, which had been inherited from the British, to names more appropriate to their own culture. Thus the streets of Victoria and Elphinstone became Abdullah Haroon and Zaibun Nisa. Only over a period of years did these names gradually come to be commonly used, for the older names died hard. The address of our office residence was 212 E.I. Lines, and although I am sure the street had a new name, I do not believe I was ever aware of it. It was always known as E.I. Lines. The area surrounding our office had once been a British Military Cantonment. The name "Lines" referred to a row of tents, barracks, or residences, while the prefix designated the occupants or honored an important personage. E.I. Lines therefore indicated that an East Indian detachment had been billeted on this street. The next street over, Staff Lines, was where the staff officers would have been found, whereas Jacob Lines was no doubt named after General Jacob.

In July 1977, the military took over the government and put an end to the civil disturbances, thus allowing Union Texas to make plans to put a seismic crew in the field. A Western Geophysical vibroseis crew was mobilized from Dubai to shoot a reconnaissance program utilizing all of the roads and tracks on the block. Realizing that this would be a seismic problem area, Union Texas and Cities sent a team of seismologists who were acquisition specialists to the field, to make sure the program was started using the optimum field parameters. This group included John Cox from Union Texas, along with Alan Shepard and Herb Drushell of Cities Service. They selected, almost at random, a stretch of line where they could begin test operations.

By the time this exploration program had begun, I had replaced Bob Stover as General Manager in Singapore. In addition, I was put in charge of the overall Pakistan operation. Elgin Deidrick, previously a staff geologist in our international group in Houston, was appointed Resident Manager in Karachi, and he reported to me in Singapore. I followed the progress of the exploration program very closely and I excitedly grasped any new bit of information as it became available. When I saw the first seismic test line that had been shot, I could hardly believe what it showed. Although it was not the anticlinal rollover I had hoped for, it was at least second best—a large, tilted fault block. To have found this structure with the short test line meant either that we were very lucky or that the area was full of such features. Thanks to improved seismic technology over the years and to the superior ability of our geophysicists, we were seeing the true structure of this part of the Lower Indus basin for the first time.

It later turned out that we drilled our fifth well on vibration point 34 of this test line, making the first discovery of oil in the Sind with the Khaskeli-1 well, but a lot happened before that.

The Indus alluvial plain in the Sind is cut by an extensive system of canals that are generally oriented north–south. In one sense, this helped out in the vibroseis program, but in another it caused considerable problems. Lines could easily be run along the canals because rudimentary roads had been built from the dredging spoils, thus allowing access to the vibrator trucks. Going in an east–west direction was tough, however, because the canals had to be crossed and only a few bridges existed. Even so, we had few serious problems and the 2500-kilometer (1553-mile) reconnaissance program was completed in 19 months. This survey gave us a pretty good idea of the structural grain of the block and we felt confident we could find a good prospect for our first well.

The first seismic lines to be processed showed us why the previous, older generation seismic had not been able to define Cretaceous structure. Even with the state-of-the-art technology we were currently able to employ, it was difficult to remove the first multiple of the basalt reflector. That was frustrating. This multiple often coincided with the reflection from the top sand of the Lower Goru formation, which we thought would be the most likely reservoir objective. In the areas of low structural relief, it would be very difficult to separate the indistinct bona fide reflections from the multiple, and we could appreciate the problem of trying to map using the seismic data from 20 years earlier. Of course, in the areas of strong dip in the Mesozoic, picking the reflections would be easier because there would be considerable divergence between the dip of the multiples of the flat-lying Tertiary and the reflections from below the unconformity.

To me, the most surprising result of the initial seismic was that the gravity maximums I had supposed were highs were shown in fact to be structural lows. In my original concept, I had anticipated that these large gravity features represented anticlines. We found out that there was an inverse density gradient because the Upper Goru had a high carbonate content and was denser than the Lower Goru. The gravity highs then represented thicker sections of Upper Goru preserved in the structural lows beneath the Tertiary/Cretaceous unconformity. This was something of a disappointment because the gravity maximums represented the principal structural features on the block, and we learned that they reflected lows rather than highs. When we had shot a tighter seismic grid, that was exactly the picture that emerged—large, closed structural lows with a few small closed highs. Perhaps I had been remiss in not realizing this earlier. In looking back at it, I imagine that Stanvac must have been aware of this relationship.

The Concession Agreement required that we spud a well within 24 months or drop the acreage. When we got to the point of making that decision, we unfortunately did not yet have a firm prospect. To some degree, the Stanvac wells told us approximately in which area we should expect a favorable sand/shale ratio in the Lower Goru formation, but we had not yet found a good structure to test. No anticlines with well-defined, four-way dip had been found, and the fault block we saw on the seismic test line had not yet been detailed. Nevertheless, we decided to continue with the concession so we chose a location near the town of Badin, which was structurally located on a broad high with a rather poorly defined south dip. Our maps had been based on the Jurassic limestone reflector, which was the strongest reflector in the section after the Deccan basalt, even though it was far below the objective of the Lower Goru sands. The well, Patar-1, spudded in January 1979. It had no shows through the Lower Goru, but near the base of the Sembar formations we did find a thin sand that appeared to be oil bearing. We wanted to test this sand, more for fluid information than the hope of commerciality, but a mechanical problem in setting the liner precluded running the test. We had to be satisfied with just having seen the show.

The second well was drilled on a tilted fault block structure considerably better defined than that on which the Patar well had been drilled. The second well, the Tarai-1, had a few shows of oil at the base of the basalt but nothing of significance in either the Lower Goru or the Sembar formations. While drilling the top of the Jurassic limestone, however, there had been a show of gas that, when later tested on an RFT, recovered gas and condensate. That was the first recovery of hydrocarbons from a well in the Badin area. When sidewall cores from the Lower Goru sands were taken for routine lithological studies, oil staining was seen. Further evaluation was needed. On subsequent testing we recovered some oil in the mud, and the tool test chamber, which captured a sample of the last fluid passing through it, contained clean, greenish oil. It was certainly unusual crude oil—light gravity and without a trace of gas. The drillers assured us that no oil had been added to the mud, and from all appearances it had come from the formation.

We sent a sample of the oil to a lab in Singapore, and a chromatograph showed it to be diesel. What were we to think? Sometime later, our geologist, Bob Pile, came across the drilling report for the day they had drilled the section we later tested: "Stuck pipe, spotted fifty barrels diesel oil." That diesel oil had entered the formation and stayed there to be produced clean on the drill-stem test. None of us had ever seen that before.

In visiting Karachi during that period, we frequently had problems with hotel space. The Hotel Inter-Continental, which was the best in town, didn't always have a room when one arrived at the customary hour of 3 o'clock in the morning. They were good at phoning around to find you a room in a second or third best hotel, however, which is how one night one of the Cities Service people ended up in the Beach Luxury Hotel. The Beach Luxury was neither on a beach (it was on a muddy creek bordering a mangrove marsh) nor was it luxurious. Although it was not a hardship staying at this hotel, our associate was taken aback when he returned to Singapore and his management, being fooled by the name, hesitated to approve his expense account.

From time to time I had similar experiences of arriving at the Inter-Continental and finding no room. Once, however, I arrived and did have a room, in which I unpacked and went to bed. I was almost asleep when the phone rang. It was reception to say they had put me in the wrong room and to ask that I please repack because they were sending a bellboy to put me in my proper place. I was somewhat surprised, because I had stayed in almost every room in that hotel and had never noticed much to set them apart. I told the receptionist this, asking what was so special about the other room. "Mr. Young," he said, "the other room has a basket of fruit in it." I suggested they bring the fruit to me in the morning.

At about that time I moved to Karachi to take over as manager. I had been making frequent trips to Karachi from Singapore, but I very much wanted to be there full-time myself. Jesse McCollum agreed to the move, although he had some misgivings because he feared he might not have a place for me elsewhere if the Pakistan project proved unsuccessful.

As our exploration program continued we got more seismic information, which gave us a much clearer picture of the regional structure. Basically, it appeared to be a horst and graben structural province with a northwest–southeast trend. This was intersected by a fault system that ran north–south. The horsts and grabens may have been part of a failed rift associated with the Cambay graben in India. There certainly weren't the anticlines I had expected from the gravity information.

During our study of the seismic in the southwestern part of the block, we noticed a reflector that seemed to be discordant with all the others. We traced it around on the seismic and were able to map a strange feature that we at first thought was a submarine canyon cutting into the Cretaceous section. It was shallowest and narrowest to the north and widened and deepened to the south. Upward, it terminated at the Tertiary/Cretaceous unconformity at a depth of about 450 meters (1476 feet); its base neared a depth greater than 4500 meters (14,760 feet) as it plunged towards the Rann of Kutch and the Indian border. If this were a submarine canyon of Cretaceous age, it seemed to me to be running in the wrong direction. I had no other idea as to what it could be.

Les Beddoes, with Cities Service in Singapore, suggested that I give a paper at a Southeast Asian conference, on "A Submarine Canyon in the Cretaceous of the Indus Delta." I declined because I did not feel confident that we knew what we were looking at. Not until the drilling of the Jati well did we find out what it was.

Before we started on our second drilling program, civil disturbances again occurred. Religious fanatics made an attack in Mecca, and for some reason the Americans were blamed. The U.S. embassy in Islamabad was burned and the American consulate in Karachi was harassed by a mob. Our office was not far from the consulate, so we soon knew something was happening when tear gas floated in on the breeze. Our first action was to take our company sign off the gate, thereby quickly transforming our office into an ordinary Pakistani residence. We had to send all our staff home because it was impossible to work through the tears. The next day our home office instructed us to send our expatriate dependents to Singapore until the troubles were over. Four months passed before we had our families back home in Karachi again.

The drilling program we then started was for three wells. The first was located deeper into the basin from the Tarai well, where we expected to find less sand and perhaps more mature source rock. The second location was far into the western part of the block, where the structure looked different and perhaps showed a stratigraphic trap associated with the questionable submarine canyon. The third location was not yet agreed upon.

The first well in this program, the Damiri-1, had a lot of gas shows in sandstones possessing little porosity. Their permeability was so low that the rates on test were insignificant. The second well, the Jati-1, gave us something of a surprise. We got the answer to what our "submarine canyon" really was.

Local camel breeders visiting the Damiri-1 well site.

When the drilling of the Jati-1 approached the seismic horizon we had identified as a canyon, we hoped to find some sands filling in an incised valley. We didn't encounter sands, however, and when we drilled the reflector we found it to be an igneous intrusion—a dolerite sill. There were traces of oil both above and below this sill, indicating that the hot intrusion had baked oil out of the adjacent source rock. These were the only shows in the well, which was later abandoned before reaching the deepest objective because of mechanical problems.

Reexamining our interpretation, we were fascinated to trace the route of this intrusive body. From great depths in the south, where the melt must have originated, we could see how it had progressed upward in the section until it reached the surface and sourced the great flows of the Deccan basalt. In places, the sill could be traced as it ran along the sedimentary bedding planes for some distance. Upon reaching a fault, it would either rise up the fault plane until it reached another incompetent bedding surface, where it would then run horizontally again, or it would cut straight across to whatever bed was juxtaposed on the opposite side. Perhaps there is material here for a technical paper, but it would have to be from the viewpoint of igneous petrology rather than petroleum geology.

The Jati-1 well was abandoned without finding anything significant, and attention was shifted to the next well of the program. We had several locations to choose from at that time, and we felt that if this well was dry, it would be difficult to continue our exploration program.

After considerable discussion in a meeting with our partners, it was decided to drill the structure we first saw on the test line.

Bill Wood, the Manager of Operations, and I met in the field with Mukhtar Ahmed, our Field Administration Officer, to see what problems we would run into in building a location and access road to the selected area. That this was a somewhat isolated part of the block was one of the reasons this structure had not been drilled earlier in the program. Bill was always a bit unhappy when I arranged for us to meet Mukhtar in the field, as I would arrange to rendezvous by reference to the seismic lines we had shot. For instance, I would tell Mukhtar to meet us at the intersection of line 12 and line 3. Although Bill knew the field area rather well, he didn't know the layout of the seismic lines at all. Being a person who always wanted to know exactly what was going on, he was rather annoyed.

I always enjoyed meeting Mukhtar in the field because he would bring with him a lunch of curry and chapattis, which is a flat, unleavened bread. We would sit on our heels in the shade on the banks of a canal and watch the local boys swimming or washing their water buffalo. Eating by scooping food up in a small piece of bread takes as much practice as learning to eat with chopsticks, but food always tastes better when one eats like a local.

On this particular reconnaissance trip to the Khaskeli location, we began to worry that we would have a real problem in building the access road. Mukhtar drove for miles cross-country, fording canals and going through swamps. We wondered if it was at all practical to drill in that area. When we finally arrived at the location, we saw that it was on a nice, level, dry piece of unoccupied ground. But what of the road? We questioned Mukhtar about alternate routes. He said yes, there were other ways, and we could take one on the way back. We were amazed to drive four or five miles along a canal path and then switch onto the main east–west blacktop road that crossed the block, with no problem at all. Mukhtar had just wanted to show us how difficult access could be so we would appreciate his cleverness in finding a simple route.

The Khaskeli-1 well was spudded in late March 1981. When drilling had almost reached the top sand of the Lower Goru formation, I drove out to the location to be at the scene when the drill hit the objective. I remember that I was sitting in the drilling supervisor's trailer reading a book when I noticed a sudden increase in the tempo of the squeaks from the brake, indicating that the rate of penetration had picked up—which usually happens when you drill into a sand. I was up on the rig floor as quickly as I could get there. The driller had stopped drilling as a safety precaution, and he shut the mud pumps down so he could check to see if the well was flowing. There was no flow, so he started to circulate mud again and we waited for the rock cuttings to make their way from a depth of almost 1000 meters (3280 feet) and arrive at the surface. When they did, Sayeed Ahmed, the well-site geologist, took them into his trailer lab for a quick check. There was no doubt, we had found oil. We could tell from both the visible oil stain on the sand grains and the detection of gas entrained with the mud, that in all probability we had a well.

The oil shows continued for a good part of the day and it appeared that we had several hundred feet of pay. This promised to be a large field. Several days later, however, when we ran electric logs we were to find out there was only 50 feet of pay. The remaining shows were all residual oil from below the oil/water contact.

That evening as we drove back to Karachi, the sun was setting behind the hills of the Pab range, west of the city. I thought back to the first visit I had made to Pakistan when I had climbed those hills to see the outcrops. That had started me on the road of exploration of the Lower Indus. Now, 16 years later, I was returning from the well site where we had just discovered the first oil in the Sind.

Although the Khaskeli field did not prove to be nearly as large a field as we had been searching for, it had the redeeming feature of shallow depth and high production rates. The total reserves of the field proved to be only about 10 million barrels, a small fraction of the size we thought would be the minimum to be commercial. Nevertheless, by designing a simple flow station and trucking the oil to the Karachi refinery, the 5000 BOPD field producing rate was sufficient to be commercial. Within the ten years that followed this first discovery, more than 30 oil and gas discoveries were made on the Badin block and the adjacent acreage operated by OGDC. Oil production from the area reached 35,000 BOPD. Renewed exploration throughout Pakistan was responsible for increasing the daily average production from 9000 BOPD in 1977 to 63,000 BOPD in 1991.

Western Canada

Miracles at Elmworth

By John A. Masters

> *...the incredulity of mankind, who do not truly believe in anything new until they have had experience of it.*
> —Niccolo Machiavelli, 1537

The discovery of the Elmworth gas field in Canada is the story of how a huge field was found in an area considered to be extremely unfavorable for the discovery of oil or gas. The area had already been evaluated and abandoned by virtually every large company in the business. The discovery was made using a method that had no recognition in the industry and that conformed to a geologic concept that was unknown in the industry at the time.

After the discovery, all the experts said we were crazy. Our production tests, which had flowed up to 10 million cubic feet of gas per day (MMCFD), were discounted in the conviction that no significant accumulation could occur in an area previously disqualified by so many major companies. All manner of negative speculations were advanced authoritatively by the pipe smokers and other experts as they explained why their previous conceptions were still right, why their status as non-acreage holders in the area was justified, and why there was no reason for them to take any disturbing action to correct anything.

New ideas are extraordinarily difficult for the human mind, trained or untrained, to accept. "There are none so blind as those who will not see." The net result was that proponents of Elmworth were a very lonely, very isolated, small group. In the beginning, it was just I; gradually it grew to include a few other explorationists in Canadian Hunter. The situation exemplifies the fact that a new idea, almost by definition, is first held by a minority of only one. No one else has it. If the idea is supported by a lot of people, you will have that warm sense of the herd about you, but the idea won't be new. An idea has to be out there all by itself to be new and, hence, to have great potential.

I don't mean to claim a hero's medal, but my serious counsel to anyone who aspires to conceive a breakthrough idea is: be tough, keep your confidence, and expect rejection and criticism—there is nothing so resented as a new idea. More people have new ideas than have the moral courage to stay with them or the good sense to exploit them and make them work. More counsel: by your loneliness you may judge the distance you have come. Unfortunately, a stupid idea will often bring forth the same amount of rejection, so it is difficult for a thoughtful person to avoid anxiety. All this means that the world does not release its bounty painlessly.

START-UP

The adventure started in 1973, when Jim Gray and I left Kerr-McGee. Well, we were asked to leave. I had managed the Calgary office for 6 years with indifferent results, but I was beginning to sense an opportunity in tight gas sands. Mr. McGee wanted me to come back to Oklahoma City and take a new role in our Gulf of Mexico exploration. I loved Calgary, I believed I could find gas in Canada, and I didn't want to go to the head office. I had been there. So, I told Jim I was going to try to talk McGee into letting me stay in Calgary but, if necessary, I'd quit. He said, "You'll never quit. McGee won't let you." I didn't think he would, either.

I went to Oklahoma City and had a discussion with my old friend and admired mentor, Dean McGee. He really did want me down there. I told him how much I wanted to stay in Calgary. After a lot of back and forth, he said, "Well, if you want to stay that much, I guess you'd better, but you'll have to do it by yourself." After 20 years! Then he said, "What do you think Jim will do?" I wasn't ready for that. My first impulse was to lie—but I realized I couldn't do that. Still, I mumbled and hesitated, and finally got out, "Well, Mr. McGee, I don't really know—I suppose—well, I'm pretty sure—he'll probably go with me." Mr. McGee said, "I thought he would. You can't keep a couple of broncos fenced in forever. You remember, I did the same thing once."

We had a few more parting words, very cordial ones, and I got out of there. With the first phone I found, I called Jim.

"Jim, I've got good news and bad news for you."

"First, what's the good news?"

"I quit."

"Oh, fantastic, I didn't think you'd have the guts."

Pause. "What's the bad news?"

"You quit, too!"

Ever since then, Jim and I have spoken interchangeably for each other. What one says, the other may not think is perfect, but we adapt.

We were a little startled at how soon we were separated from Kerr-McGee. It seems a couple of people were delighted to see us go and

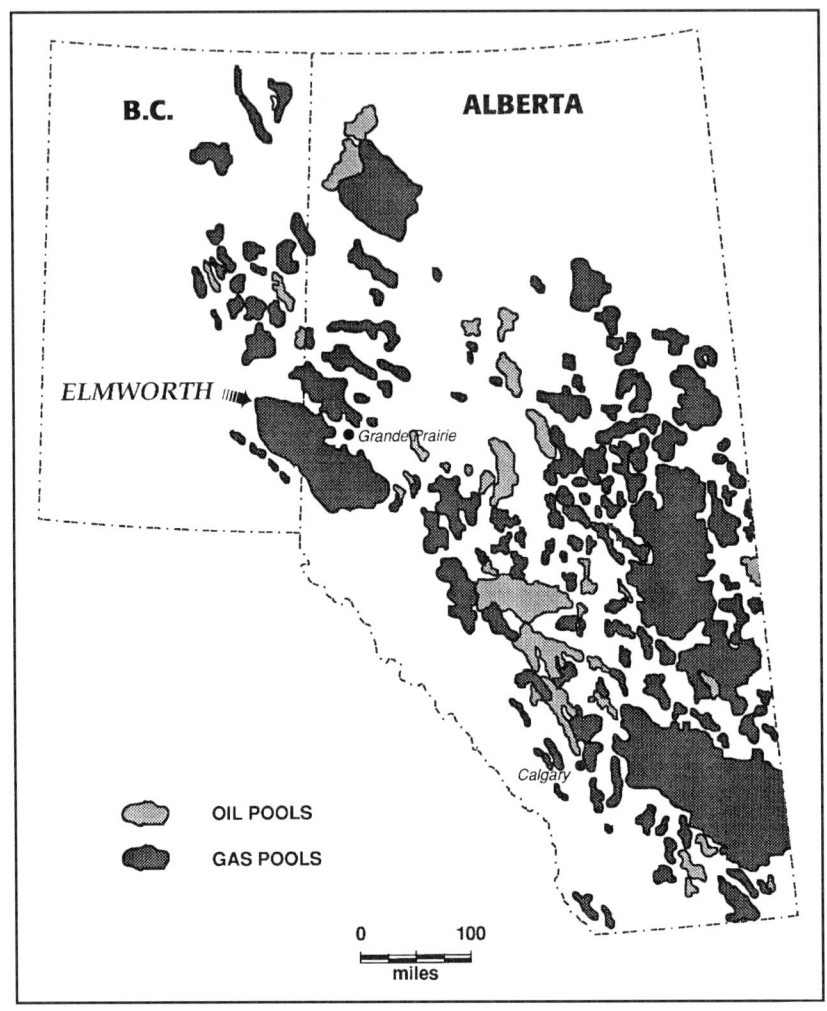

Location map of Elmworth field.

they saw to it that in a few days we were history. You know, most people prepare for their exodus with job interviews, a new position, and liberal use of the xerox machine, all designed to make the big leap as safe and painless as changing apartments. But we went out on the streets very abruptly. We had almost no money. Both of us had been on pretty small salaries all of our careers, and had very little saved up in the bank. We began our new lives in a borrowed office: one desk, two chairs, and a telephone.

In those days, when Jim and I were trying to start Canadian Hunter Exploration, we had charts showing our price projections. We had the gall to predict that someday gas would reach 60 cents per thousand cubic feet (MCF) and that oil might eventually go to $6. Everyone thought we were hopeless.

After being turned down by 12 major industrial companies in Canada, who, we hope, are now tortured by regret, we made a financial arrangement with Noranda Mines and started our new company. Well, actually, we made a tentative agreement with Alf Powis and Bill Row, President and Executive Vice President, respectively. We made it because Alf had decided that "When everyone else is leaving an area and you see a good reason to stay, that's what I call opportunity."

I remember the morning when the phone rang at home and I was still asleep. Ozzie Hines, Powis's executive assistant said, "Well, I guess we've got a deal. You guys may prefer to wait for a full legal contract, but that will take a couple of months. Otherwise, I can deposit $100,000 in the bank and we can get started on a handshake." It took me about 10 seconds to choose the handshake and the $100,000.

Three years after Jim and I borrowed that desk, we found the largest gas field in the history of Canada, at Elmworth. MIRACLE. Now, at the beginning of the 1990s, we employ 375 people and are the 13th largest gas producer in Canada. We like to think that the good guys won, after all, but the issue was seriously in doubt for some years.

CONCEPT

I left Kerr-McGee with an intuition that Canada probably had large reserves of low-porosity sandstone gas, which had not yet been exploited. I remembered from my U.S. experience that much development was going on in the U.S. in porosities as low as 5%. In Canada, most of the gas reservoirs were above 15% porosity, and the very lowest was 13%. Most Canadian geologists, believe it or not, used to explain this by saying there weren't any sandstones in Canada with less than 13% porosity; or they said that lower porosity sands would never produce economically. Yes, that's what they said then, although they will probably deny it now. It would make an interesting chronicle to record the number of misconceptions in the oil industry, by prominent, well-informed people, that were later proven wrong by experience (like "I'll drink all the oil west of the Mississippi" by a Standard Oil executive). Such a chronicle would serve as a caution about today's conceptions, which we all think are so immutable. However, back to my story. In spite of conventional opinion, I knew there were plenty of low-porosity sands in Canada and that new hydraulic fracture techniques in use in the United States would make low-porosity sands produce. Jim and I were absolutely convinced that gas prices would rise and the advanced technology would become economic.

I began my work with Canadian Hunter by examining several different aspects of gas accumulation in western Canada, not knowing for certain whether any of the investigations would fit together or whether all of them might be futile. I built a map of gas shows across the whole region, which was twice the size of Texas, that recorded the test recoveries of gas in abandoned wells. I carefully noted the formation tested, flows from each test, pressure data, and the like. I also began to construct a group of 500-kilometer- (300-mile-) long, east–west electric log cross-sections across the basin from eastern Alberta to the foothills. These sections showed the entire geologic column. What I was looking for were sand changes, porosity changes, anything that seemed anomalous or curious. At the same time, we made a deal with a small consulting group from Houston, called Sneider & Meckel Associates, to help us with our geology and technology. They had an electric log analyst, Lloyd Fons, in their group and they called him "the fastest gun in the West." Sneider convinced us that

In the beginning, there were two of us. Jim Gray is on the right.

there should be some number of bypassed pay zones in Canadian wells and log scans were the best way to recognize them.

Fons had worked with electric logs at Byrd-Frost, Schlumberger, and as an independent for 30 years. He could scan electric logs and recognize potential productive zones faster than any man I've ever known. He would sit at a table with a stack of logs, lay the resistivity and porosity logs side by side, run his eyes down both logs in one continuous motion and drop them into the "good" pile or the "bad" pile. Later, he would calculate the good ones. He could examine over 200 logs in a day with amazing accuracy. At night, he would do them while sitting in front of a TV screen watching basketball games.

Now here was a system that really had a lot of muscle. You see, we had 75,000 abandoned exploratory wells in Canada. This represented several billion dollars worth of geophysical information; hard information that could tell you "yes" or "no" about the presence of oil or gas; geophysical information that no explorationist had ever used to explore on a regional scale. Virtually all electric logs in the past had been used for correlation or on single wells as an evaluation tool at the conclusion of drilling. We converted a close-in, development tool to a reconnaissance exploration method. It seems now like a simple, obvious idea, but the very fact that it had not been done before posed a formidable barrier. There was a strong argument that the major companies would never have by-passed any significant pay zones. The exercise was deemed to be futile. But, many of the 75,000 wells had been drilled looking for oil at a time when gas was at a very low price, when there were practically no gas pipelines in western Canada. It is a curious fact that the human mind sees mostly what it is programmed to see. Those human minds recognized that gas was not a profitable commodity so, by and large, they did not see it. In actual fact, they were trained not to see it because the companies did not want to waste drill-stem tests and completion costs. Shell Crewson, an old-timer in Calgary, remembers his days as a well-site geologist for the Hudson's Bay company in the Elmworth area. The young geologists were told not to bother with the uphole part and were not sent to the wells until after the Cretaceous section had been drilled.

Another factor of extraordinary importance was that between 1972 and 1977, the years when we were starting Hunter, the price of gas increased nine times. Suddenly, gas had very important new economic significance. At the same time, there was an effective, new hydraulic fracturing technology available in the U.S. Billions of dollars worth of geophysical information existed in old well logs, which no one had ever used in a regionally consistent, logical exploration program. It was a time of great change but, of course, that is more apparent now than it was then.

We stood, almost unaware, on the brink of an oil company's most precious position—that of being first in the industry to use a vital, new exploration tool. If you think back over the history of the industry, you can see that the first companies to use the plane table for structure mapping, the first to use the core drill, or the gravity meter, or the seismograph, always creamed off a number of easily found discoveries. So it was with electric logs. We asked Lloyd to make a rapid exploration scan of a representative selection of all the exploratory wells in western Canada. To any normal log analyst, this would have been an absolutely impossible request. To Lloyd, it seemed like a big job, but we were paying him $1,000 a day, so he was glad to go to work on it. He was confident he could get it done. He recorded his work in big note-books without any regard to the map locations of the good wells or the bad wells.

When I took on the task of plotting Lloyd's data on a regional map, I found that he had evaluated some 5000 wells across the eastern Alberta shelf, and only 18% of those wells had indications of gas. I put big dots on all of them. Most of these wells were scattered randomly across the shelf and did not show any abnormal clustering. However, there was a group of wells in the western part of the basin that did show unusual and exciting characteristics.

Several hundred wells were in the western, downdip, deepest part of British Columbia and Alberta, in the area we now call the Deep Basin. A large proportion of them indicated significant, heretofore-bypassed gas reserves. There was an obvious eye-catching concentration of these big wells in a large area south of Grande Prairie, near the little village of Elmworth. This concentrated area of wells was 160 kilometers (100 miles) long and 80 kilometers (50 miles) wide. It contained 85 dry holes, which Lloyd evaluated as bypassed gas wells. I enclosed that area with a roughly drawn contour line that designated greater than 8 bcf of gas per section. I drew two more contour closures within that area, labelled 15 bcf and 25 bcf. These contours almost exactly delineated the outline of the great Elmworth gas field and its central core area of maximum reserves. As far as we know, it was the first time a bcf-per-section map had ever been made from electric log analyses. I made my contour interpretation of the entire map, which stretched from Saskatchewan to British Columbia and covered a whole wall, in two days. What I am saying is that, after the leg work, it took only two days to find Elmworth. MIRACLE.

But that gets ahead of the story. As the electric log map was taking shape under my hand, my mind raced for an explanation of this great concentration of gas in a synclinal position, the one place it should not be. By this time, as part of my regional studies, I had constructed east–west electric log cross sections, which did not show the sandstone pinchouts I was looking for but did show rapidly increasing electrical resistivity throughout the entire Cretaceous section as the deeper part of the basin was reached. The Cretaceous sands out on the gentle eastern shelf are clean, porous, saltwater-bearing sands with resistivities of 2–3 ohms. When a gas accumulation exists, it presents itself as a very sharp increase in resistivity. As the shelf dip increased into the Deep Basin, over the hinge line, the sands increased rapidly in resistivity. Even the shales showed high resistivity. The whole section finally exceeded 200 ohms. For many years, Canadian geologists had observed this phenomenon and said that it simply meant very tight, highly cemented rocks. I asked our log analysts to consider very carefully whether the high resistivity in the Deep Basin could mean the whole section was saturated with gas. This possibility, if it had ever occurred to others, had probably been dismissed as ridiculous. But they had never seen the San Juan basin in New Mexico, where the entire 5000-foot Upper Cretaceous section is saturated with gas.

I was beginning to think San Juan basin with regard to the Elmworth area because another set of my maps was starting to show a very interesting pattern. A map of gas shows indicated a large area on the eastern shelf that had small accumulations of gas with abundant recoveries of water, downdip. It was very common to have recoveries of both gas and water from the same test. So I had a large area of mixed gas and water on the eastern shelf. But in the Deep Basin in western Alberta, where there was an exciting concentration of important electric log shows, I also saw a number of gas tests, but no water. Underline that in your mind. There was no water with those Deep Basin gas shows. So I drew a 250-kilometer-long northwest–southeast line that separated the mixed gas-water province of the eastern shelf and the gas-only area of the Deep Basin. The gas was downdip from the water. It saturated every porous stringer over a vertical section of 2100 meters (7000 feet). That was evident in the type log. MIRACLE.

This map brought very sharply to mind one of the most striking characteristics of the San Juan basin accumulation. The gas is in the syncline at the bottom of the basin where there is no water, and grades updip through a gas–water transition zone to fully water-saturated rocks. You can draw a gas–water contact line along the updip edge of the gas. This is upside-down from the normal gas-on-top-of-water relationship, which is the only situation the geology textbooks talk about.

By that time, I had experienced that flash of intuition that suddenly sorts out numerous random, subconsciously remembered data in your

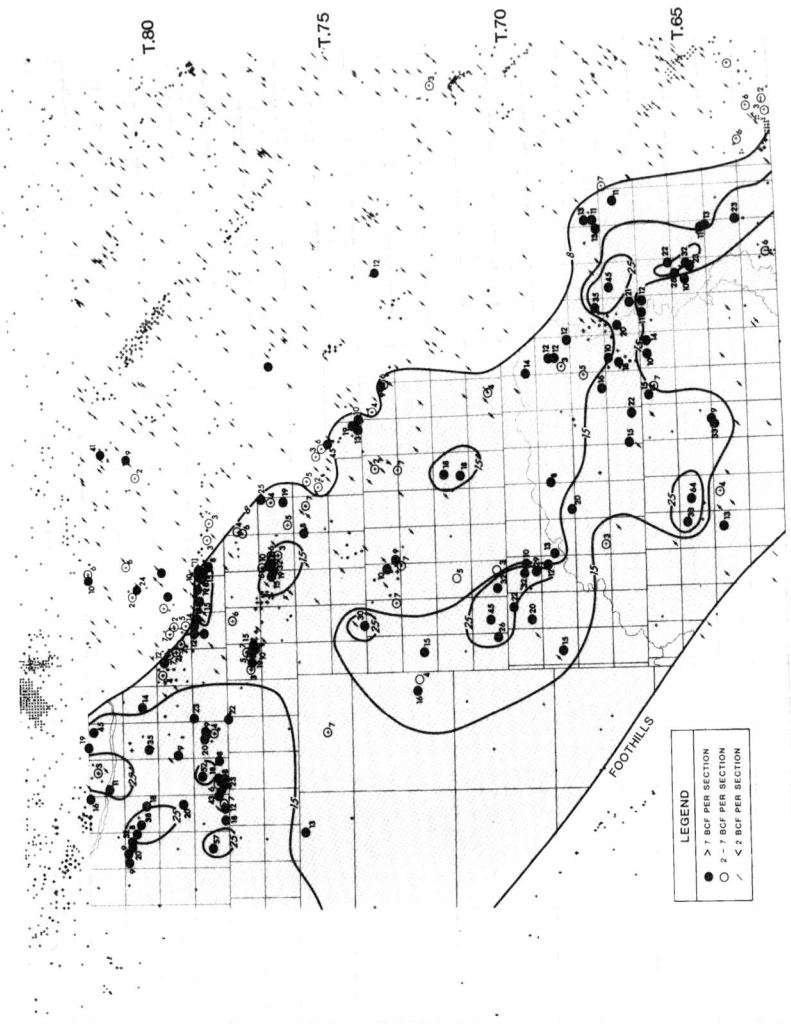

First BCF-per-section map of Elmworth.

mind into an ordered pattern. Eureka, all is revealed. The flash is accompanied by a powerful, exhilarating sense of understanding and control. You stand on a peak. You *know*. It is life's grandest feeling. It far surpasses mere pleasure because, for a time, it represents unlimited knowing. MIRACLE.

From that moment, sometime in the summer of 1975, two years

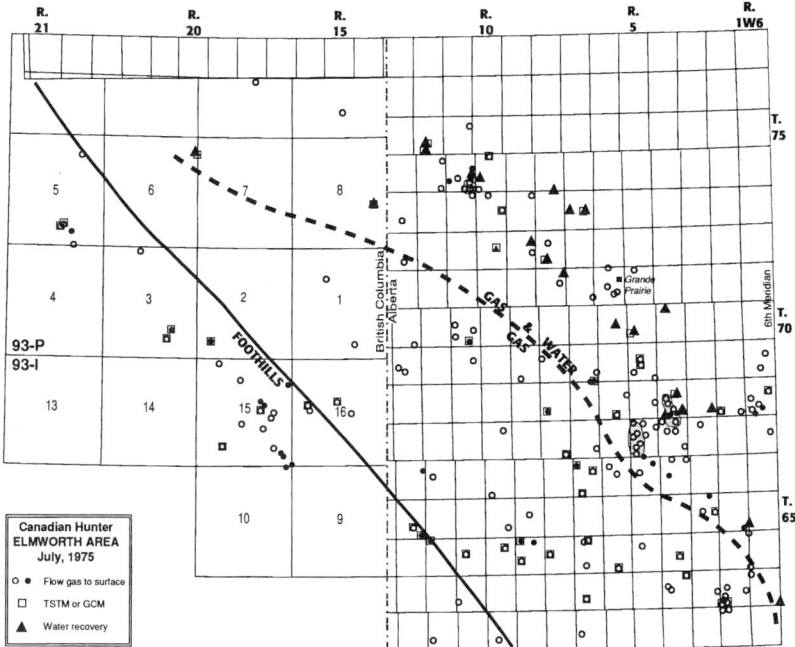

The Elmworth area.

after we had started the company, I knew I had recognized a field more than 160 kilometers (100 miles) long. I knew it would be one of the largest fields in North America. No person or institution ever shook me from that conviction. Yet no geologist, even from the San Juan basin, had ever described this inverted gas-water situation in the literature as a recognized gas trap mechanism.

My mind clicked to the conditions I had read about in the Denver basin, where R. A. Matuszczak with Amoco found the large Wattenburg field. His evidence was 22 dry holes, none of which had gas tests larger than TSTM (too small to measure). But, there was no water until you got on the updip side of the accumulation. The wells covered an area of 950 square miles. Matuszczak recognized the similarity to the San Juan basin and led Amoco into the discovery of a 1.5 tcf field, which they developed with the aid of massive hydraulic fracturing.

As I moved ahead vigorously to complete my electric log sections, numerous facies and thickness maps, gas show maps, and the like, I contacted my old friend Elliott Riggs in Farmington. "Can you write me a no-nonsense, summary account of the important facts about the

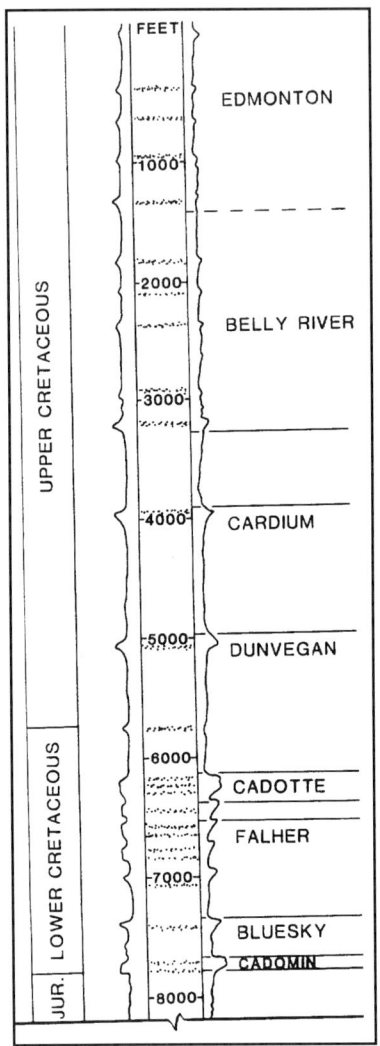

Type electric log, Elmworth.

San Juan basin gas accumulation?" I asked. He said he could, and in a few weeks I had in my hands probably the shortest and best commercial geology report ever written on the San Juan basin. For a year, we checked every observation in the Deep Basin against that report and used it to project and anticipate conditions we had not yet seen.

I had another analogy in mind that I wanted to study—the giant Milk River field in southeastern Alberta. There, I knew the Milk River sands were porous updip to the south where they were saturated with water. Downdip, the sands became shaly, had low porosity, and were saturated with gas throughout a very large area. They contained reserves of 10 tcf. At the time, Milk River was the largest gas field in Canada. San Juan is the second largest gas field in the U.S., having 25 tcf. These inverted gas traps in tight sands, downdip from water, were unknown in the textbooks. Ten years ago, no geologist accepted these traps as a recognizable concept in petroleum geology. Nevertheless, this kind of field holds some of the biggest reserves of gas in North America. The new concept, and the fact that large fields described by the concept already existed but were not consciously recognized as being different, present a precise example of how human beings, including geologists, can be blind to anything but accepted wisdom.

To further understand the intellectual climate of the time, you should know that the Canadian Geological Survey had just said in 1975, "It is likely that the larger pools ... have already been found and exploration efforts in the future will be devoted to searching for remaining smaller pools."

Most of the major companies had already reached this conclusion and moved their exploration efforts to the Arctic and the east coast. The banks, consultants, columnists, and pipeline companies all agreed. A pretty solid wall of negative thinking was in place when I reached the startling conclusion that there was a huge, San Juan basin-type field in the Elmworth area, which contained many trillions of cubic feet of gas. It might be the largest in Canada.

I was no longer alone. Jim Gray and Bob Sneider, Dave Smith, Meckel, Fons, and a few others had bought in and were believers.

ACTION

Jim and I went into Toronto and laid the huge electric log maps, which covered British Columbia and Alberta, out on the big conference table in front of the president of Noranda, Alf Powis. I said, in my most historic manner, "Alf, you are looking at the most valuable set of geophysical maps in Canada today." I went on to explain the whole geologic concept. Alf is not a geologist, but he is a quick study, and he rapidly grasped the main idea. Where he got stuck was with the 85 dry holes, which I said were bypassed gas wells. Alf said, "You're trying to tell me that a couple of dozen major companies drilled 85 wells through a huge gas field and every one of them missed it? I find that very hard to believe." Alf found it very hard to believe, and the Noranda Board of Directors found it even harder to believe. But Alf finally said that if we were convinced, he would go along with us in spite of the others.

Our first move was to get Jim Chaput to negotiate a 220-square-kilometer (50,000-acre) farm-out from Texaco around the village of Elmworth. The size of that farm-out tells you something about how unfavorably the rest of the industry regarded this area. These days you're lucky to farm out 2500 acres anywhere in Alberta. I was particularly interested in a gas show in the Cadotte formation (Viking equivalent). I wanted to reenter a well and test that zone. We did that. The zone was too tight to produce commercially. Our farm-out allowed us, we thought, to drill another well to a deeper depth. Our geologists had identified the Falher sands (below the Cadotte) as a possible secondary objective, but we didn't attach much importance to them. When our second well drilled through them, we recognized some porosity and had a gas kick, so we tested and it flowed 4 MMCFD. I call your attention to the fact that the discovery zone is not the one for which I was looking. This also happened at many other major discoveries. MIRACLE.

Elmworth is not much. The telephone was used a lot.

A few years later, Jeff Larson, a Texaco land man, said to Bob Stone, "You know, Bob, you guys defaulted on our deal. I remember the day we talked about it all afternoon and came within an ace of bouncing you out." Bob said, "I don't know what this is about." Jeff went on to remind him, "If you read that agreement letter, it says that Canadian Hunter will reenter the old Cadotte well for the purpose of attempting to complete the Cadotte. Then it says if you do complete the Cadotte and get a producing gas well, you have the right to go ahead and drill option wells. But you never did complete it. When you wrote and said you planned to drill the first option well, we had a lot of talk about disallowing you. But, we decided, 'What the hell, there's nothing there anyway!' "

It gives us cold chills to recognize that we nearly lost Elmworth before we found it. The field would surely have been discovered by someone, but no one ever would have heard of Canadian Hunter. MIRACLE.

Left to right: John Masters, Bob Sneider, Jim Gray, and Larry Meckel at the discovery well in 1976.

It seems dumb now, but our first judgment indicated that the trend was northeast. Dave Smith and Larry Meckel both went on to do a lot of fine work in the early period of Elmworth, but we used to beat up on them a lot for that first interpretation. The discovery well was actually our second well. Against all statistical probability, that well, the first successful well in Elmworth, is still the second most prolific well we have ever drilled in the field, and we have since drilled 300 wells. It is very useful to have your luck early.

Let me tell you some more facts that are statistically entirely improbable, but that happened. In this business you get used to the unlikely. We located the third well with great expectations and drilled a dry hole. The fourth well was dry. The fifth well was dry. The sixth well was dry. We were getting desperate. Noranda was getting desperate too; those damn things cost several million dollars apiece. Our seventh well was located a long 10 miles to the southeast and we hit big again. The well blasted in for 18 MMCFD from the Falher. We were ecstatic with joy and excitement. We had cores and logs. The logs correlated point for point with our discovery well. The stratigraphic trend was clearly northwesterly.

When the cores arrived in Calgary, Jim, Bob Sneider, and I went out to Corelab to examine them. The pay zone was a coarse gravel conglomerate. Jim and I were like a couple of kids picking up pieces of core, enthusing about the porosity, exclaiming about the size of the gravel pebbles and admiring the total thickness of the pay zone. But Sneider hadn't said anything. He had his hand lens out and he was examining every foot of core, very carefully, in sequence. Finally, he looked up with a big grin on his face and said, "Goddamnit, you guys, this is a beach. Every one of these zones coarsens upwards. That means they aren't river channels, they're shoreline beach sands. This conglomerate is the top of the beach. I've seen exactly the same thing in California." MIRACLE.

Jim and I knew enough not to need any further explanation. We knew that channels are narrow, curvy, and discontinuous. We knew that beaches are wide and long and straight; they can go for miles. We suddenly realized that we actually had an important discovery. A beach trend could hold very large reserves. And our two big wells, 10 miles apart, demonstrated a northwesterly trend.

Let me admit now, before you think of it, that this very important piece of geologic interpretation about beaches had to be pointed out to me. I'm not ashamed to use any good idea that's available.

From then on, things were easy. We drilled dry holes now and then, but most of our wells were gas wells. Eventually one of them was AOF (absolute open flow) tested for 340 MMCFD.

Our understanding of the beach environment was a very critical piece of geologic information. For many months the competitor companies, knowing that the pay zones were conglomerates, assumed that they were river channels. They seem to have been predisposed to a negative interpretation and they did not have enough knowledge of sedimentation to make a proper environmental interpretation. Sneider knew how to do that from his training at Shell Research under Rufus LeBlanc. That was a small piece of geologic understanding that was worth hundreds of billions of cubic feet of gas to us from successful land buying, because we understood the values better than our competitors and outbid them in sale after sale. MIRACLE.

We had other advantages. Our electric log maps showed us very clearly where the main volumes of gas lay. But of equal importance was the geologic concept of the regional trap, analogous to the San Juan basin and other big fields. We knew theoretically, academically, intellectually that all those widely scattered dry holes, which we analyzed as bypassed gas wells, were part of one big field. We were not paralyzed by the apprehension that they might be only small, isolated pools separated by large, dry areas. We knew we were in a big field. We knew it was continuous. Therefore, we believed that essentially all the acreage was good in a 165-kilometer (100-mile) strip, 80 kilometers wide, although we weren't yet ready for British Columbia. We hit the land sales like a ton of bricks, buying everything in sight. At one point, I even wrote a memo to Powis saying that maybe the most important action in his career at Noranda would be finding money for us for the next several sales because those sales would determine whether we acquired a giant natural gas asset and became a major company. Well, maybe I was a little presumptuous about Powis's career, but he has always been tolerant of my bursts of enthusiasm.

MONEY

The discovery came in January, 1976, a few months after the Geological Survey had written off western Canada. For the next 2 years, we spent money faster than Noranda had ever experienced in an exploration play, drilling wells and buying land. We had times when we were wildly enthusiastic and the Noranda management was solidly with us. Wells were coming in bang, bang, bang. And there were other times when the well results were poor and Noranda management lost confidence, largely because of outside advice from the banks, the trade journals, and even from Jack Armstrong, Chairman of the great Imperial company. At first, Jack thought our idea of a huge, continuous gas field downdip from water was nonsense. Later, he changed his tune completely and bought 12.5% interest for $175,000,000—but not before he nearly destroyed us with the Noranda Board by telling them they'd better get out of the deal before those two young fellows took them to the cleaners.

Neither did we do ourselves any good with the Noranda Board in 1977, when Alf persuaded them to give us a budget of $14,000,000 for the whole year. Two big sales came up early at Elmworth and we spent all the money by March! The Board was horrified. They were accustomed to operations that spent one-twelfth of their budget each month. Powis looks back on that as one of the most acute embarrassments we ever dealt him—in a long string of them. Now, in the long view, we can see that those purchases were the best buys we made in the whole history of Elmworth.

After our spectacular budget overrun, Noranda wouldn't come up with another dollar. It was the old authority syndrome: teach 'em a lesson. That gave my partner, Jim Gray, a chance to show his stuff. First, he made a fast deal with Sulpetro for $24,000,000 for half interest in whatever it would buy. When it ran out, he had similar deals ready with Kerr-Addison, Canadian Commercial Bank, Klaus Hebben from Petromark, Capital Bag and Grey Beverages from Vancouver (who would have thought of them?), and Carl Nickle with $1,700,000 one hour before a land sale. MIRACLE.

SWEAT

Somehow, we band-aided our way through 1977. It is only in retrospect that the money-raising problems appear difficult. At the time, we couldn't tell one miracle from another. Our total staff of 50 people were also performing miracles with oil invert mud, new frac techniques, new log and DST (drill-stem test) interpretations, new sample examination methods, gas agreements covering the largest area Trans Canada had ever contracted, designing and building the biggest throughput gas plant in Canada (the first plant we had ever built), and a whole lot more. I repeat, 50 people. MIRACLE.

The early days at Elmworth were chaotic. During one of the early winter drilling periods, Hunter had 21 rigs running along with 10 service rigs. In 1980, Jim Boyd, our completions supervisor, took reports and gave operating instructions to all 10 service rigs for 340 consecutive nights. During the days, he designed the fracs. He had a calculator but no secretary, of course. We were stretched, sure, but we were excited, too. I don't remember any complaints. The unanimous attitude was, "What can I do to help?" MIRACLE.

Jim Boyd has a claim to genius. Almost single-handedly he invented the fracture procedures at Elmworth. He learned how to frac conglomerates by being right there in the control trucks playing the wells to improve their flowback efficiency. He showed the industry that you can take good wells making 10 MMCFD and hammer them up to 25 MMCFD. No one had ever done that before as standard operating practice. (The rule had been, "If it's a good well, don't mess with it".)

Our drillers accomplished their share of miracles. The 21 rigs were directed during the winter of 1979 by three drilling engineers and two supervisors. Drilling costs started out at $4,500,000 in today's dollars; we are now down to $700,000 for a drill-and-evaluate job. It took a lot of brains and experience to do that, and a lot of hard work on rig floors in the bitter cold.

DISBELIEF

Despite our land successes and our wells, those were some of the toughest times, when we were being scorned by most of the industry and most of the experts. I can scarcely remember anyone who agreed with us. Perhaps the most remarkable thing was how little that shook our confidence. We were true believers. We knew we were right! The criticism bounced off us like bullets off a tank. What I remember clearly, however, is a powerful sense of loneliness and a fear that by the slightest possibility we actually might be wrong. What if the field actually would not perform in long-term production? We might have spent all that money in vain. Our careers would surely have been ruined. No one would ever trust us again. But I'm proud to say that we didn't waiver.

The mind-set of some government geologists can be illustrated by my vivid memory of an event that took place in our offices in early 1977. It was important to us, with regard to Noranda and other investors, to encourage some favorable opinion among authorities in order to point to reputable scientific thinking that agreed with us. To that end, we invited the director of oil and gas reserves assessment for the Canadian Geological Survey, and his assistant, to listen to an all-day presentation on the geology and reserve potential of Elmworth. About 4 P.M., at the end of the day, everyone was tired, the room was relatively quiet, and the director turned to his assistant, covered his mouth with his hand, and whispered in a louder voice than he intended: "I sure hope they're not right. It would mean a lot of work for us."

We had a similar experience with the National Energy Board (N.E.B.) engineers. Realizing that they were doubtful of our claim, we invited them to a private presentation in our offices. Several of us described the geology and engineering data we had, which, if known publicly, could have led anyone to hundreds of millions of dollars worth of gas. One of the most enthusiastic detractors on the board dozed off during my talk. None of the others showed any interest. No one asked any questions. We had tried to show them the facts of our discovery. Unfortunately, they appeared to be completely disinterested.

We took our story to the British Columbia Department of Energy in Victoria. After a long afternoon presentation, we were thanked by the Minister, the Chief Geologist, and others. Finally, the Chief Engineer, an old Shell hand, said, "All I'll say is, I never heard so much bullshit in one afternoon in my life."

The Geological Survey and the N.E.B. had taken such definitive public positions claiming no more large fields would be found in western Canada that they appear in hindsight almost to have had a vested interest in not recognizing them. Elmworth had become a hundred miles long and still the N.E.B. refused to believe it. It is curious, and a little frightening, that they could be oblivious to the data for so long. By 1979, three years after discovery, the N.E.B. had stretched to the recognition of

1 tcf of reserves. Eleven years later, the field had produced nearly 2 tcf and was producing at a rate of 550 MMCFD, which equals 1 tcf every 5 years. You may be inclined to excuse them, but remember, this is the field that is now producing into the pipelines 6% of Canada's total gas and gas liquids production.

My Elmworth experiences put me cheek by jowl to those in our industry generally considered to be the "experts." The media thinks so and the individuals do, too. They have comfortable, well-paying jobs with banks, investment houses, government regulatory boards and research groups, universities, and the like. I agree that they are smart people; they can all write competent reports. But, unfortunately, they are bewildered by anything new.

Some people in the industry were beginning to see the light, however. Certainly, all the companies who were competing with us at the sales understood that the Deep Basin was a big gas field. In June 1978, I presented a paper on the field to a large conference in Calgary. Jim Law is a Calgary consulting geologist. He said he had never before been in a meeting where he could describe the geologists as "thunderstruck." In 1979, the paper was published in the *AAPG Bulletin*.

In July 1978, Carl Nickle of the *Daily Oil Bulletin* wrote a special editorial:

> In our view, there is no longer any doubt but that the Deep Basin concept outlined by John Masters is valid. In fact, the Canadian Hunter team deserves the primary credit for establishing through research and huge risk-capital spending what may well be one of the largest Natural Gas Reserve Basins in the world.

In August 1978, Imperial changed its mind and decided that there really were big fields left to find in western Canada and Elmworth was one of them. With Jack Armstrong's blessing, Imperial's Vice President of Exploration offered us $60,000,000 in drilling for a range of interests in the field. After just enough financial analysis to convince us they might pay more, we counter-offered and finally settled on $175,000,000 for 12.5% interest. It was the richest farm-out deal in the history of Canadian oil, up to that time. MIRACLE.

That deal confirmed us to the Noranda Board. Mark Millard, our wonderful friend from Loeb Rhoades, exulted to them, "This is a miracle, because, when the deserving get what they deserve in this day and age, that is a miracle. Gentlemen, Canadian Hunter's reputation is now absolutely BULLET PROOF. Their reputation has been attested by the largest oil company in the world."

That was the turn of the tide. There were still carpers and naysayers, but they were being left behind, stranded on the flats, sinking into the mud. We watched them with little sympathy as they went down

When all the geologists in town started to see straight again, Canadian Hunter, a tiny little company, virtually unknown to the industry, had sailed in under the noses of 400 other companies and found a monster, the biggest gas field in the history of Canadian exploration. We estimate the total potential reserves in the Elmworth field to be about 16 tcf. Canadian Hunter has about one-third of that, making us an important gas producer. We found it $2^1/_2$ years after we started our little company. MIRACLE.

INSIGHT

Elmworth was only the second exploration experience in which I was so intimately familiar with the geology, the land, and the players that I knew what was going to happen. I had done this once before, in the exploration for uranium at Ambrosia Lake in New Mexico in 1955.

Jim Chaput, head land man. He has steered every deal we ever made.

I knew the feeling of complete intimacy, that sense of transcendence when you are no longer bound by the usual risk odds, ordinary factual data from wells speak to you with uncanny meaning, the purpose and probable action of other companies is prophetically visible.

For a period of time, perhaps the 3 years from 1977 through 1979, I knew what other companies would bid. I knew this because of our wells, scouted information from competitor wells, a keen sense of how the play was developing, what importance and perceptions a recent snippet of new geologic information would have, and so on. There were land sales almost every month. Values were escalating at every sale. Many times, after exhaustive consideration of sale bids in conference with all our experts, I would have some change of heart, call Jim Chaput the night before, and change bids up or down, sometimes doubling them. Now here is a fact that should catch your attention: on 83 sale parcels, throughout three years, we won 82! We bought 3000 square kilometers (675,000 gross acres) for $130,000,000, averaging $200 per acre. This established the bulk of our land position in Alberta. I don't hold much with E.S.P. and don't cite it here. All I call attention to is that you can improve your performance to a level well above normal if you really focus on something, live and breathe it night and day, and get in a position where you have more information than anyone else. We worked hard at that last element, recognizing what piece of land was a critical outpost that would give us a vital drilling location for obtaining the most forward drilling-coring-logging-testing information to evaluate a future sale. We made sure we won the bid, then we moved a rig on immediately. We posted land very aggressively, trying to keep the competition off balance. We aspired to Admiral King's dictum: "Hit 'em and keep on hitting 'em." When we tested a new pay, or found one going in a new direction, we bought land immediately. And we bought it all. No horsing around with testing the market.

LAND

We commonly are pestered by ill-informed people. I am amused by the geologists who tell me they recognized Elmworth many years before I did but their managements refused to believe it. However, I resent the criticism that we were successful bidders because we threw money at the sales. The record shows that we rather carefully rode the crest of increasing values, staying just enough ahead of industry to nearly always win. Yet our prices never reached uneconomic values.

Following are notes from my files, summarizing the early sales:
- December 1975. 220 sq. km. (50,000 acre) farm-out from Texaco. Net earning cost $30 per acre.
- March 1977. Bought 1200 sq. km. (280,000 acres) in two sales at $50 per acre for $14,000,000—all we had.
- December 1977. Another 1260 sq. km. (290,000 acres) at $116 per acre for $34,000,000. (Prices had doubled in 9 months.)
- December 1978. 365 sq. km. (84,000 acres) at $275 per acre for $23,000,000. (Another doubling).
- June 1979. 580 sq. km. (135,000 acres) at $385 per acre, $50,000,000. (Price had increased 11 times since the beginning.)

I know of no other major land-buying campaign in public lease sales throughout a 3-year period, in which one company so dominated the market. Land men have been confessing for years to Jim Chaput that on Parcel X we beat them by only $5.00 per acre, on Parcel Y it was only $1.75. MIRACLES.

Let's go back now to the early days of the field, so you can see that others were as confused as we were. In 1975, Texaco had the extraordinary good fortune of making two simultaneous farm-outs in the Elmworth area, both of which were excellent discoveries: the Canadian Hunter well in Cretaceous conglomerates, and 33 kilometers (20 miles) to the north a Chieftain well, which found 95 feet of Triassic sand and tested 27 MMCFD in the Sinclair prospect. The acreage had been bought originally for Devonian potential, drilling results had been discouraging, and the land was then being held for the usual reasons that support the usual policy never to drop land. At the same time, Texaco was building a huge refinery in Ontario and was strapped for cash. As new land came up for sale around the discoveries, Texaco was forced to choose where to put its limited dollars. At that time, they badly needed an expert in sedimentary environments for they had to evaluate which of the new wells was in the reservoir that would be most extensive and dependable. They chose the Triassic incorrectly because they interpreted the Cretaceous conglomerates to be northeast-trending stream channels.

We went to their offices for a bid meeting prior to a sale. (We always went there, because it wasn't seemly for Texaco to come to our humble premises.) We would name a price, they would suggest half of it, we'd discuss, they would mourn that Toronto would never approve, finally we'd propose they take some proportion of the bid, less than half. As a result, they ended up with shares like 40%, 25%, down to nothing. This is the first public explanation of a trivia question that has bothered land men for many years, namely, why did Texaco and Hunter have such varying interests in the numerous Crown blocks they bought together in the Elmworth area.

One time, we didn't decide on the bid for the following day until late in the afternoon of the day before the sale. The bank had closed. One of our guys phoned Mac Alston, Manager of the Canadian Imperial Bank of Commerce. Mac said, "How much do you need?"

"Twenty-five million."

"Good Lord, you guys, that's me, personally. If the auditors are here tomorrow, I lose my job. I can't do this. You want 25 million, and you haven't got a hundred bucks in the bank!"

We pleaded. Mac wouldn't budge. So we got Jim Gray to call him.

"Mac, what the hell's wrong with that cheap outfit you've got over there? What's the big deal about 25 million bucks? You know we're good for it."

Mac came through. MIRACLE.

Our stories about land sales would fill a book. Imagine, we had three guys in the land department who spent $130,000,000 at 29 sales in

Most of our drilling was in the deep cold, because of muskeg.

three years. I don't know how many individual checks, and calls to the bank, and trips to the sale location, as well as bid meetings and discussions with partners, were involved. Think of the most awful goof-up imaginable, and we did it. But, like the "Perils of Pauline," the mistake was corrected in the nick of time. Some of these near catastrophes were so horrible they are just being admitted to me, 10 years later, as I gather material for this book. One time, I find, Gary Aitken left approximately $21,000,000 in certified checks in an envelope on his kitchen table and caught a plane for the sale in Victoria. Glen King was contacted and immediately solved the problem by catching a direct flight and beating Aitken out to Victoria.

Scout reports on competitor wells were vitally important to us in assessing the proper bids at sales. Before Elmworth, scouting had not been done with the dead seriousness with which we practiced it. The ethic seemed to be: "You can't steal in Calgary, but anything goes in the field." Telephone lines, mobile radios, roughnecks in a bar, whatever you could make out through binoculars, mud men, DST crews—all these were fair game. Our own people learned never to call from a pay phone or a motel phone, unless they were sending deliberately misleading information. We comforted ourselves that we only stole from the rich.

Scouting in the Canadian bush has its own peculiar problems. One time, a scout called our Calgary office from his mobile. The first thing he said was, "Wait a second, I've got to roll up the window."

"What's the matter? Is it cold?"

"Naw, just a bear trying to get in here with me."

Cold? Sometimes it was 50° below zero.

Another combat trick we developed at Elmworth was to buy most of our land in brokers' names. While all the controversy was going on and Canadian Hunter was being pilloried as a fraud and a promoter, we seemed to fit that definition because we didn't appear to be a serious competitor for the land. It was being bought in the names of a variety of different land companies, agents, etc. No company had ever covered their tracks that way over a large area, over a considerable period of time. When we finally announced via the *Daily Oil Bulletin* that during a period of 3 years we had bought more than 600,000 acres in an area 160 kilometers (100 miles) long by 80 kilometers wide, the message was sent to everyone that we were serious. Since Elmworth, most companies have adopted our tactics of hiding the real ownership of bids, if they are going to be involved in a sequence of sales.

Jim Chaput had to live with comments from other land men like, "Hey Jim, how come you're at all the sales and you never buy anything? Canadian Hunter is all talk!"

If our approach to competition at Elmworth sounds sometimes a little too opportunistic to be entirely fair, think of it as guerilla war. Our company was very small, and our budget quite limited. We were buying land at bargain prices in the biggest gas field in Canada, scared to death that one or more majors would wake up and blow us out of the water. I learned what a burglar feels like creeping through a house at night. We didn't have the money or the muscle to play the game straight up. We had to use all the smoke and mirrors we could find. MIRACLE.

FLAVOR

Let this story describe how we patched together our operation. It took a lot of iron to supply 20 drilling rigs, so we hired a Manager of Purchasing, Gerald Schultz, age 27. After two weeks on the job, he placed an order for $27,000,000 of pipe. Then he thought about purchasing policies at his last job, so he hurried up to Jim Gray.

"Jim, I've got a question. Does it bother you that I just signed purchase orders for $27,000,000?"

"No."

"Do I need to have these approved by anyone else?"

"No. I don't know anything about pipe, and I doubt if anyone else does. Isn't that what you do?"

"Yup—I guess."

"Well, go ahead and sign 'em."

"What if there's a problem?"

"It's your ass, not mine. Is that fair?"

"That's fair," said Gerald. "I can live with that."

Schultz remembers getting to know Jim Boyd, our completions man, who had run more than a thousand fracs in his 17 years at Halliburton. Jim said, "Let's have a meeting." That meant the discussion had turned formal. "You know those valves that have the handles like this?"

"Yup, WKMs."

"I like those. Let's not buy anything else."

That is how Canadian Hunter's valve policy was set.

Behind this apparently whimsical and free-wheeling operations procedure stood a very sophisticated engineering management. Jay Christensen was Vice President, Production for several years and then stepped down voluntarily (but is still with us) for Al Dillabough, who came over from Shell where he had been manager of the largest volume of gas production in the industry. We hired only the best. Some people dubbed Canadian Hunter the "Canadian Rustlers."

Elmworth has been a triumph of engineering as well as geology. In fact, all the functions of an oil and gas company were successfully integrated into one of the industry's most closely meshed operations. The gas field is a tribute to many people and many disciplines, but that is another story for another time.

PROOF

Everyone in the industry was infected, to some extent, with skepticism about our wells. In part, we encouraged this by continuing to talk publicly about "tight sands," but never uttering a word about our big Falher conglomerate wells. Whenever we had a potential investor, someone from Noranda, or whomever we needed to impress, we made a quick plane trip to Elmworth, got in a helicopter, flew over our vast field, reminded everyone they had never seen anything so big, pointed out the 20–30 rigs drilling at any one time, and then landed at one of our big gas wells. This was what came to be known as "John

Elmworth—for as far as the eye can see.

and Jim's sacred ceremony." A testing crew would be on site with the well hooked up to a flare line directed into a big pit with a high dirt back wall. When all were safely gathered behind the well, my partner would give a John Wayne signal like "Let 'er go, boys," and they would slowly crank the well open. The gas would ignite with a WHUMP! The tester would continue to crank, calling off the flow numbers from his gauge, "Five million ... eight million ... ten million ... twelve ... fourteen ... SIXTEEN MILLION ... [now shouting at the top of his lungs] IS THAT ENOUGH, MR. MASTERS?" He was coached to indicate unmistakably that he could keep going, but everyone would be at serious risk. The ground was shaking, the noise was like a 727 taking off, the heat was ferocious (we had everyone standing so close they would have to back away), you couldn't hear someone yelling into your ear. After several minutes—long enough to make people think that the world was on fire—I waved for a shut off and the tester cranked it down quickly to a lazy flame and then it was gone. Silence. The dead, unearthly silence of the northern bush, 580 kilometers (350 miles) north of Calgary. You could hear one or two people cough, then a nervous laugh, then suddenly one of the visitors with a big smile would clap a Hunter on the back. Someone would walk over and shake my hand. Then, pandemonium, everyone joyful, everyone excited. Most of them had never seen Hell tapped into. The scene never failed to have an extraordinary impact, particularly on people who had had a sneaking suspicion the day was going to confirm some of their worst doubts. I would gather everyone around and tell them they had witnessed a miracle. I would say, "You have just seen the only practical solar energy system you will ever see. That gas came from the energy of the sun, 100 million years ago, that was transformed by photosynthesis and concentrated into a jillion tons of plant life. It was later buried and chemically altered, by the heat of the earth, into methane gas. I wanted you to come out and see that miracle. And just as an added little touch, we wanted to show you that it is not all stored in tight sands. We have some conglomerate reservoirs about 2100 meters (7000 feet) below us which are among the highest quality pay zones in all of Canada. A lot of people don't believe this, but now you can say to them, 'Have you ever been up there and seen those wells?' " That was always followed by a cheer. Each time, we took home a happy group of people. Few of them would doubt any longer.

Events have confirmed the confidence we tried to impart at those ceremonies and belied the evident showmanship. There are now more than 1000 wells in the whole Elmworth field area. Canadian Hunter owns 300 of them. Average spacing is 10 wells per township, so there is an enormous amount of infill drilling left to be done. Some day, far in the future, we will see thousands more wells there. Average Absolute Open Flow potential of all wells is 9 MMCFD. The largest AOF is 364 MMCFD and the largest deliverability at 1000 psi is 135 MMCFD. These wells drain Canadian Hunter's proved and probable reserves of 3 tcf. Large amounts of potential reserve await measurement. Years into the future we will still be fracing the tight sands that were our original vision of the field.

BRITISH COLUMBIA

But, there is more to the story of the whole field. In 1977, our drilling had reached the British Columbia border on the west end of our Elmworth field. Already, the field was 160 kilometers (100 miles) long. We thought it likely that the beaches would extend across British Columbia to the mountains, but what if the trend changed angle? A great deal of land was involved and there were few wells across British Columbia to give us guidance. I've forgotten who thought of the idea to look at the outcrop along the mountains, but if we could find the telltale beach conglomerate, with continental sediments to the south and foreshore

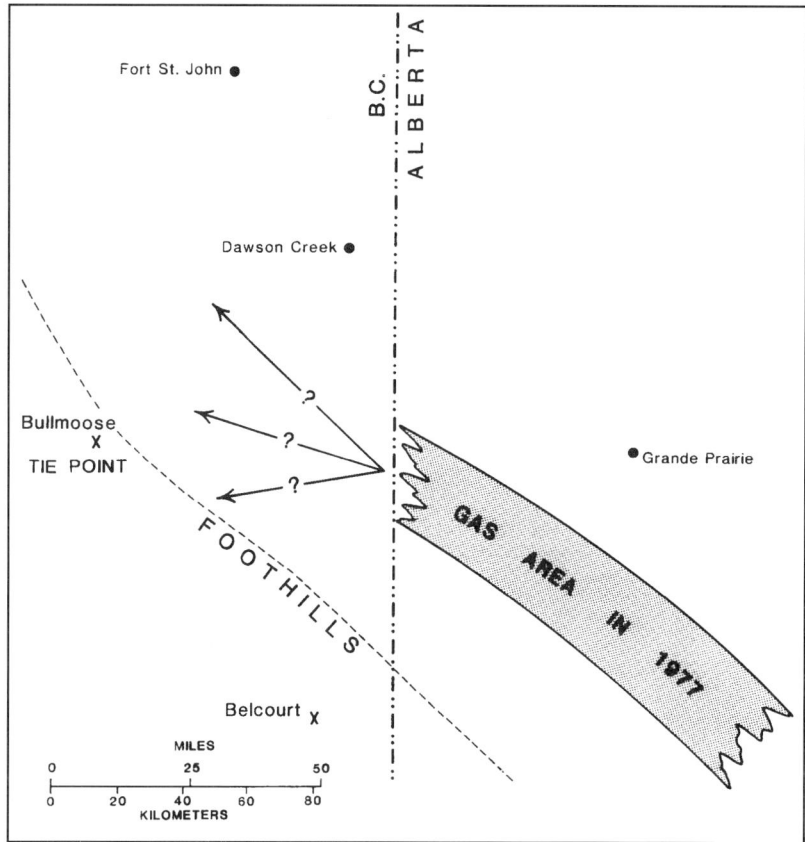

The play across British Columbia.

and basin sediments to the north, we could nail the shoreline. Then we could project the trend from our last control well in Alberta all the way to the outcrop location.

Larry Meckel, our smart sedimentary environments consultant, offered to make the trip. In fact, we would have had to beat him off with a stick. Dave Smith went with him. They flew to Fort St. John and took an Okanagan helicopter across virgin bush for 160 kilometers (100 miles) to the belt of outcrop along the east side of the mountains.

They landed first on Mount Belcourt, got out, ducked under the blades, scrambled up the cliff, looked carefully, hammered the rock, and examined it with a hand lens. No good. Coaly shales, lenticular sands. Continental. Got in again. Flew a few miles north. Landed again.

"You could hear the goddamn seagulls!"

Same thing. Did it again. Same thing. Did it again. And again. And again. Mid-afternoon. Hot. Discouraged. Maybe the shoreline had turned north. Maybe we would never see it in the outcrop belt. Let's land one more time on those ledges down there. Bull Moose Mountain. Glory to God! Rounded, sorted, coarsening-upward, stratified beach conglomerate, 10 meters thick on top of fine-grained foreshore sandstones. Two separate cliffs of conglomerate. We had the tie point. Meckel and Smith were wild, like hockey players after a goal. Over there in Alberta, seeing those rocks in cores and samples, the shoreline was an intellectual concept. Here, in solid rock strung out for miles along the cliffs, it was real.

Meckel called me from the first telephone he could get to when they landed in Ft. St. John. He shouted, "John, we found it!"

I said, "Are you sure it was beach?" He said, his voice lowering to seriousness, a sentence that has become part of Canadian Hunter mythology: "John, you could hear the goddamn seagulls!" He heard them across 100,000,000 years of time. MIRACLE.

Well, we had it then. We laid the beach trends across northeastern British Columbia for 115 kilometers (70 miles) to our tie point at Bullmoose Mountain and started posting land. We went to four sales in British Columbia during the next year and bought 1150 square kilometers (265,000 acres). Then, the story gets sticky. We drilled one dry hole after another. We just couldn't find enough permeability after we crossed the British Columbia border. In total, we drilled 24 wells before we gave up, licking our wounds.

Six years went by and we continued to drill and frac and learn in Alberta. In 1984, a little light turned on in my head, and I said, "It seems to me we've learned so much in the last few years about evaluation and completion that we make wells out of holes we would have abandoned a while ago." Agreement from all. "OK, then it's time we went back and took another careful look at all those B.C. wells we drilled."

I selected Gay Jervey, a very bright geologist we had hired from Esso, and told her she could choose anyone she wanted for her British Columbia team. "I want you to have the best." Gay chose carefully and assembled a team made up one each of a stratigrapher, an electric log analyst, a petrographer, a DST analyst, a hydrodynamics specialist, and a completion engineer. This team evaluated very rigorously all the wells previously drilled by Hunter and other companies. After intense examination, Gay reported in a final review that many of the wells would have made commercial completions with current methods and the team had measured several tcf of potential gas reserves. MIRACLE.

We identified more than 1,000,000 acres of favorable open land and planned an initial drilling program of 20 wells that would cost over $1,000,000 each. We were going to need a trainload of money—too much for Noranda to risk on one prospect. Because Canada is a prolific, underexplored area, and we are good explorers, it makes business sense for us to find the big prospects and farm out part interest to others for the front-end risk money. That way, we get to test a number of prospects and retain a substantial interest in all of them. That doesn't make as much sense for a company that has a hard time finding good prospects.

About that time, we had a call from BP asking if we had any producing gas properties for sale. We said, "Yes, we have a large gas property we'd be interested in talking to you about." Being just a tiny bit dishonest is sometimes best for everyone. Jim and I were invited to Houston to make our pitch to the whole exploration management team of BP America. We presented our maps and logs and snake oil story to Dick Bray and a whole herd of grimfaced, humorless note takers. We told them we had a discovered, indicated gas reserve of 6 tcf and they could have half of it if they would put up a very large amount of money. Bray's reaction at the end of the day was that we had been quite persuasive about the geology but, of course, the money was crazy. Still, he assigned one of his handymen to the job of negotiating a satisfactory deal. We got to know Preston Rennie pretty well over the next couple of months as we haggled over every bcf and every dollar. Many times, I was ready to walk away from the Chinese water torture Rennie used as a negotiating method, but my partner, Jim Gray, who sees good in everyone, and believes every dollar spends the same way regardless of the per-

Gay Jervey, team leader in British Columbia.

son from whom you get it, persisted. It was agonizing, but we finally reached agreement on a set of options that would ultimately allow them to earn 50% interest by spending $400,000,000. Not too shabby. It is the largest farm-out deal ever made in Canada and it was made largely on acreage that still belonged to the Crown. The last negotiations were with Bill Johnson, who had taken Bray's place, and we signed the deal in Cleveland. Alf Powis came across the lake from Toronto to comfort BP that there really was a responsible corporate entity who could pay their share when the time came. After signing, I said to Johnson, "Well, now we've got to find the gas." He replied, "Don't worry about that. Those Falher beaches are going to go right across. If we'd had any concern about that, we'd never have joined." Later, Powis said to me, "You guys must be the world's greatest salesmen. Nothing is that sure." Both of them turned out to be right. The Falher beaches, which had been confirmed by the seagulls, did go across but were generally tightly cemented. The overlying Cadotte beach conglomerates came in and were beautifully porous right at the British Columbia border and saved our skin. To give credit to Gay, that is exactly what she expected. MIRACLE.

Over the next 5 years, we bought 3900 sq. km. (900,000 acres) for $97,000,000 and drilled 135 wells, largely with BP's money. Proved reserves in these irregular sands are laborious and expensive to measure by reservoir engineering methodology, so it takes time and money to do it right. But we judge that there are potential recoverable reserves of 2 tcf indicated by our drilling on the joint lands. We are presently producing, through the first plant at Noel, 150 MMCFD, mostly from Cadotte reservoirs. Remember, Cadotte was the objective in my first well in Alberta.

PERSPECTIVE

It is an interesting insight into the thinking of many commercial geologists to realize that it took the bulk of the industry 2 to 3 years after the initial discovery to recognize and accept that Elmworth was a major field. For a very long time, geologists protected their conventional wisdom by simply denying the field existed. They could not come to grips with a large, synclinal field in an area of 85 previous dry holes. They adjusted to this by clinging to a claim that Canadian Hunter people were promoters, that we were exaggerating our well tests, the pay zones did not correlate and, therefore, were erratic and small in volume, and so on. In general, the media parroted the same story; the banks were unanimously resistant to the idea that we had anything of value; and all government spokesmen, particularly the

National Energy Board, were consistently negative. We even had the distinction of being singled out in the 1979 National Energy Board annual report for "misleading the general public" by claiming a major gas find. We suffered those slings and arrows in silence because all that time we were buying land at Crown sales for bargain prices. It suited our purpose just fine for everybody to be dumping on Elmworth, because that kept the competition down. We ended up with approximately 7000 square kilometers (1,600,000 gross acres) over the entire 150-mile-long field.

Today, at last, great quantities of gas flow into the pipelines from Elmworth, and all the doubters have long since busied themselves with casting their illumination on other projects.

Notwithstanding the various problems of the past, Deep Basin is now a name and a concept that is an accepted part of the petroleum industry in the world. Furthermore, Elmworth stands historically as the turning point in Canadian industry–government's perception of the Western Basin as still being the vital resource base for the country. A senior vice president of Albert Gas Trunk Lines said, "Masters and Gray—are largely responsible for changing the industry's thinking from deficit to surplus." Elmworth brought the major companies back from the frontier and focused exploration efforts once again in western Canada. In the late 1970s, the largest company in Canada, Imperial Oil, had decided to give up their entire land base in Alberta and British Columbia and concentrate on the ice-ridden waters off Canada's northern and eastern coasts. Their entry back into western Canada was signalled by the purchase from Canadian Hunter of an interest in Elmworth as previously described.

LESSONS

What are the lessons a young explorationist can learn from the story of Elmworth? I think the same elements run through this series of events that run through many other large discoveries.

First, if it's a really new idea, you will be a lonely minority of one. If you can't stick it out and take all the flak, you won't make it. Perhaps you will find some comfort in remembering what Edward R. Murrow said, "The obscure we see eventually; the completely obvious, it seems, takes longer." I will never forget that after discovering Canada's largest gas field, for 3 years we were the "town shits." It was open season on the rotten guys who claimed to have done something that no one else thought was possible. Now, that's all behind us, and we're among the fashionable—but, man, we had to pay our dues. You will never be able to develop a new idea unless you have the courage to accept unpopularity in the face of almost unanimous disapproval.

Second, you have to have a good idea and it has to be right.

Third, you have to conceive your idea early. It doesn't have to be scientifically precise, it just has to be right enough that you can take action quickly and decisively. Someone else is bound to think of the same idea pretty soon, because it is based on accumulated knowledge.

The intellectual achievement is only part of the battle. Don't expect anyone to have as much confidence in your new idea as you have. You must lead and maintain the momentum. The early stages present the most opportunity for audacity because only a few will comprehend your purpose.

And don't expect it to be easy. Remember, we had lots of dry holes and tough days. If you can swing it, get the kind of partner with whom you'd choose to go tiger shooting. A person you can really count on. That's the sort of person Jim Gray is.

The idea for a large, new field must almost certainly be a broad, regional concept that fits all the geology vertically and laterally into a single, consistent, integrated picture. I hold that the really effective explorationist is rarely a specialist in anything. He can seldom paint any part of the picture with the skill of a specialist, but he has the temperament and mind-set to try to understand everything. He will integrate information and data from many specialists having a wide range of understanding into one big, consistent, connected picture. His mind may never meet the test of the academician. He will not write many papers, and he will not be known for research. But his is the rarest and most valuable of minds in the commercial world. He is the conceptualizer and the integrator. He is the finder out there in the Great Alone.

Be kind to him.

EPILOG

After 18 years, Elmworth has produced nearly 2 trillion cubic feet (tcf) of gas, has plant throughput capacity of 1.2 billion cubic feet per day, and in 1990 produced 550 million cubic feet per day. Elmworth is the second largest producing field in Canada and represents 6% of total Canadian production. The field is now 140 miles long and 35 miles wide.

Elmworth's proved plus probable reserves, including past production, are 5.6 tcf. The field's ultimate potential measured by Canadian Hunter is 16 tcf.

There are 1500 completed wells in the field. Total expenditures in the field for drilling, facilities, land, plant, and geophysics are now $3.5 billion.

Truly, a giant was born.

North Yemen

Yemen Oil Hunt—
A Gift from the Gods

By Ray Fairchild

As to the possibility of a profitable (oil) concession being available near the Yemen border, I think it most unlikely. It is desperate country once away from the Asir and Yemen highlands and I should not think it likely to be oil-bearing. Moreover I cannot see any company running a pipeline through the Yemen.—

[From a letter dated 15 October, 1952 by British Ambassador in Jedda, W. N. Hugh-Jones, to R. C. Blackman, Esq., British Foreign Office, London]

BACKGROUND

So it was, always has been, and surely always shall be! Such statements and their deliverers should not be treated lightly, and certainly should not be ridiculed because of the countless instances in which they are proven to have been, or will be, so wrong! Perhaps one should consider the possibility that such seemingly endless negativeness somehow causes an ultimate metamorphosis of conditions to the extent that exact opposite conditions result! Such a transformation must surely be as good, if not more scientifically acceptable, than the generally promulgated view that all such learned and respected negative observers throughout the history of mankind were, forgive the word, wrong!

Whatever one's background, i.e., historical, psychological, managerial, or other (religion excepted), and regardless of one's reason for having chosen (if there was a choice) such a background, intentional success in petroleum exploration, in the real world, requires (1) a basic knowledge of geology as it relates to the origin and entrapment of hydrocarbons, the shape of the earth, and historical/geographical relationships therein/on; (2) respect and appreciation for, as well as understanding of, the cultural and political influences that prevail in any chosen target area; and finally, (3) an open mind that is imaginative and unfettered by prejudice, be it technical or otherwise.

It was with 30 years of personal background, including $6^1/_2$ years with Maersk Oil and Gas in Copenhagen, that I discussed international exploration with Ray Hunt in November, 1979. I recall that we were mutually surprised and pleased to learn that so many of our views about attractive areas and methods were very similar, such as (1) don't follow the crowd, in fact, go in the opposite direction; (2) don't hitchhike on others' success; and (3) don't hunt success with a scatter gun—use a sniper's rifle. Our visit led to my employment with Hunt Oil Company as Manager–International Exploration, with the responsibility of getting the company involved internationally, and, hopefully, with profitable success.

As of 20 January, 1980, Hunt Oil had a small office in London to handle the company's interest in Beatrice field in the U.K. sector of the North Sea. My immediate and primary task was to find an experienced and effective leader for that office and to increase the office's area of responsibility to include Europe, Africa, and the Middle East.

On one of my many trips to London that summer, I contacted Dr. Ian Maycock, the London manager for Zapata Exploration. I was immediately and thoroughly impressed with Ian's knowledge, especially of petroleum exploration in Europe, the North Sea, and the Middle East, where he had worked from 1964 to 1968 with Conoco. In addition to Ian's technical expertise, I found his manner and Scottish wit to be a pleasant balance to our typical Trinity River Bottom ways, and I immediately offered him a position in our organization. As I recall, it was around the 21st of July, 1980 when Ian joined our team as its third member, thus adding a significant dimension to our international effort. (Bette Winter had earlier joined our embryonic group as a multilingual secretary. She was much more talented than I at keeping things in order.)

NETWORKING

On 13 August, 1980, in a London pub near the original Hunt office on Haymarket, Ian introduced me to his longtime friend from their days with Conoco, Moujib Al-Malazi. Moujib had most recently served as Vice President of Conoco Chad before resigning to become an independent oil finder. I found Moujib to be a quiet man with sparkling eyes and explosive laughter as he told us of his ideas regarding an historic area in North Yemen near Marib, the ancient capital city of the Sabean kingdom (queendom) of the Queen of Sheba. He felt that it exhibited some evidence of a graben in association with the Red Sea

Temple remains from Sabean (Sheba) era.

and Gulf of Aden. We were intrigued with his idea, because if it were found to be true, or possible, it would qualify as the type of play we had programmed as our primary target. In support of the potential merit of the area, Ian recalled having seen significant Jurassic salt in stratigraphic tests drilled in the southwestern part of the Rub-Al Khali. Moujib assured us he would be in touch upon returning from his next visit to Yemen.

GOOD NEWS & MOVING

On 15 October, 1980, Moujib reported to our London office that he had found substantial evidence of a sedimentary basin in the Marib area.

Ian immediately undertook the design of an exploration program that would quickly and effectively verify or condemn the area's petroleum potential. His program proposal was received in Dallas on 24 October along with an analysis of what the consequences of a successful operation might be. E.R.C., a British consulting firm in London, did an unbelievable job in projecting the "consequences" of a discovery, not only with respect to volumes of oil, but perhaps even more so in terms of the design and cost of pipeline construction and other considerations in the unlikely event of success!

GO EAST, YOUNG MEN!

Plans and arrangements were then considered to facilitate our next move, which involved a visit by Hunt Oil representatives to Sana'a, the capital city of North Yemen, as well as to the basin target area.

Moujib, in the role of Hunt's emissary, conveyed our interest in visiting the target area and in meeting the Yemeni officials with whom discussions and negotiations would be undertaken in the event that all went well. We were impressed with the promptness in which invitations and visas were issued and our delegation lost no time in arriving in Sana'a on 8 January, 1981.

Hunt representatives, other than Moujib, were our Administrative Vice President, Tom Meurer, and Ian Maycock. As first-time visitors, Tom and Ian were "bowled over" by the scenery, beginning with the landing approach to the Sana'a airport. The plane began its descent as it crossed the Red Sea, passing over the coastal mountains, terraced for centuries, before touching down at an elevation of 2,400 meters (7,880 feet).

In Dallas, we were most anxious to hear that all was going very well with our intrepid trio. You can well imagine my concern upon receiving a phone call from Tom Meurer informing me that he was calling from a local jail where they had been taken by a young soldier who discovered that Maycock did not have his passport and visa on his person. Hoping this might be as bad as it could get, I sent the following telex to Tom believing that it would be read, appreciated and properly distributed by Yemen's efficient security folks:

To: Thomas Meurer, Ian May-   cock, Moujib Al-

Remnant of the Marib dam that made possible the "Land of Milk and Honey."

    Malazi
        Hunt Oil Company
        Delegation
        c/o the Sheba Hotel

Fr: Hunt Oil Company, Dallas
Date: 9 January 1981
Re: Investigation of Petroleum Opportunities in Yemen

I was pleased to receive your telephone report reflecting the positive and helpful spirit of cooperation exhibited by the Yemen representatives with whom you have met. Such an attitude is, of course, important in our overall consideration of the venture.
We hope that during the remaining days of your visit you will be able to acquire substantial information concerning:
1. History of petroleum exploration
2. Inventory of technical data
3. Field geological conditions
4. General infrastructure
5. Economic conditions and how these would relate to exploration and production
6. Concession format, and termsthat would apply
7. Petroleum laws

8. Overall assessment of opportunity in Yemen Look forward to your return no later than you indicated via telephone so that we can begin to make decisions on this venture.

R. E. Fairchild
Ref: BW 9.1.81

Somehow or other, they were released, and after an extremely busy and effective few days, they returned to Dallas in a state of near euphoria about the country, its people, and the potential for big oil in the Marib/Al Jawf basin.

BACK HOME

Assembling of the International Exploration staff in Dallas continued, with several significant additions. Jay Fortun, an attorney who had recently returned from a tour of duty with Aramco in Saudi Arabia, joined as International Negotiator; Chris Stone, a young British geologist whom I had known while in Copenhagen and who had also worked for Ian as a consultant in London, came aboard as our geologist; and Jerry Daniels, an exceptional multi-skilled cartographer, joined us after service with the United States Geological Survey and the Alabama Geological Survey.

Main gate in Sana'a.

A terraced hillside in Yemen.

GETTING SERIOUS

In London, drafting of a Production Sharing Agreement began immediately, and, thanks to the experienced guiding hand of Armen Sahakian, was completed, then negotiated with and accepted by Yemen Oil and Mineral Resource Company (Yominco) on the 28th of July, 1981. Translations to Arabic and executive approvals were then granted and formal signing of the contract took place in Sana'a on the 4th of September, 1981. Parliamentary approval of the contract was granted by North Yemen on 16 January, 1982 (approximately one year after our delegation's first visit to the area).

Initial discussions with geophysical contractors had begun on 19 January, 1981, and actual seismic operations by Geosource (Petty-Ray) commenced six days before parliamentary approval of the Production Sharing Agreement. Although that would not have been a wise move in most new areas, in this instance, as with so many events in North Yemen, there seemed to have been an agreement amongst the Gods to forgive our indiscretions, perhaps in keeping with their pre-ordination of success!

The September 1981 signing for oil exploration in the Yemen Arab Republic. Standing, from left: Ian Maycock, Ali Jabre Alawi, and Moujib Al-Malazi, consultants. Seated, from left: Ray Fairchild and Abdul Rahman Al-Bahr, Minister of State and Chairman of Yominco.

> **Production Sharing Agreement Negotiated from 22 to 28 July 1981 by the following:**
> **Representing Yominco:**
>
> H. E. Abdul Rahman Al-Bahr, Minister of State and Chairman of Yominco
> Mr. Ali Jabre Alawi, Director General, Geological Survey Board, Yominco
> Mr. Abdul-Aziz Al-Makhlafi, Director General, Planning Department, Yominco
> Mr. Saif Haza'a, Director General, Legal Department, Yominco.
> Mr. Abdulla Sallam Naji, Assistant Director General, Geological Survey
>
> **Representing Hunt Oil Company:**
>
> Dr. Ian Maycock, Head of Delegation, London
> Dr. Armen Sahakian, Negotiations Consultant, London
> Mr. Moujib Al-Malazi, Exploration Consultant, London
> Mr. Robert W. Shytles, International Attorney, Dallas
> Mr. Jay B. Fortun, International Negotiator, Dallas
> Mr. M. E. Methvin, Tax Attorney, Dallas
>
> **Executed in Sana'a on 4 September 1981 by:**
>
> H. E. Abdul Rahman Al-Bahr, Minister of State and Chairman of Yominco
> R. E. Fairchild, Vice President-International, Hunt Oil Company

Surface geological studies began on the 13th of October, 1982 under the leadership of Dr. Gerhard Martin of Bad Nauheim, West Germany. Gerhard is one of those very rare types who have been blessed with 3-D vision, a visual field of 360°, and depth perception that permits observation of effectively buried outcrops that are essential to reconstructing the geology of the area to which the person is directed. Best of all, Gerhard reports everything you really want to know, on a single typed page. We were also extremely fortunate in having Mike Morton join Gerhard for the survey. Mike had been one of the first modern-day petroleum geologists in the general area, and in addition to his knowledge of the area's geology, he provided much valuable information concerning its history and culture.

Handsome countryside native with ever-present jambya.

Dr. Gerhard Martin and his surface geology crew at tea time.

The Alif 1 discovery well.

Gerhard and Mike's work progressed smoothly and quickly, although it was not without occasional incidents during their "technical invasion" of such a remote and ancient corner of the world. For example, on a visit to a reported surface exposure of salt at Safir dome, they were fired upon by riflemen at the dome as they drove toward the feature. They stopped immediately and took cover, and were soon approached and informed that although they were welcome, their vehicle was not, since it "might contaminate the salt." The mine, located along a one-time principal caravan route, had been in operation for centuries, and those in charge, although in a rather remote corner of the world, were intensely aware of the necessity of at least attempting to maintain immaculate environmental conditions.

Other than the obvious cultural and historical interest in their visit, Gerhard and Mike discovered a very promising oil shale exposure that was so rich it was used locally as fuel, much as we burn coal. It should go without saying that this team, assisted by Yominco technicians, Asker Ali Hussein Al Tahiri, Nagueeb Abdulgalil, and their driver Yahyah, provided a wealth of invaluable information to assist in our eventual understanding of the area!

Back in London, Ian and Moujib happily engaged in that somewhat mystical yet essential game of "If this is how it is now, how did it get this way? How did eons of events affect the possibility of hydrocarbon generation, and if generated then, where are they today, if they exist at all?" We not infrequently recall the many confusing, stimulating, and sometimes inspiring hours we spent poring over data in our small office overlooking General Eisenhower's World War II headquarters near Shepherd Market.

THIS COULD BE THE START OF SOMETHING BIG!

Upon completion of the initial 1845 km (1146 miles) of seismic, there was no question that we were really onto something! A number and variety of prospects or leads were apparent, and it was clear to everyone that drilling at least one exploratory well was required. Hunt's operations departments became involved with those heavy iron problems for which they are so well qualified. Logistics, with a capital "L," became the name of their game. Location for the initial test had not yet been selected, but we knew as much as could be known at that stage about proposed total depth and the anticipated stratigraphic section.

The search for suitable contractors for the drilling phase progressed smoothly, and we were most fortunate in obtaining the services of excellent companies, equipment, and personnel. Westburne Drilling Company's Rig #212 and related equipment were landed at the humid Red Sea port city of Hodeidah. With excellent cooperation from Yominco authorities, the rig and equipment were trucked 500 km (310 miles) over the highest mountains in the Arabian Peninsula, the highest peak of which reaches 3759 meters (12,333 feet), to a dry desert environment northeast of Marib.

LITTLE THINGS MEAN A LOT

Selecting a location for our first well involved a last-minute change from the initial choice on the Lam feature to the drill site at Alif. Lam was subsequently drilled and found to be gas-bearing, which was certainly preferable to salt water, but one can only wonder how the effort might have gone had Lam been drilled first rather than Alif. In any event, the Alif location was staked and prepared, and the rig was moved in, rigged up, and spudded on 31 January, 1984.

Mosque with prayer tower.

FINDING A PARTNER

It had been decided that we would invite participation by a "partner" company to assist in defraying costs of the project through its initial phase. Numerous companies were invited to join and technical presentations were made; however, for one reason or another, none chose to participate. I then suggested to Ray Hunt that I contact Sam Frazier, an old friend, who at the time was serving as advisor to Pedco, the South Korean national oil company. I would enquire whether any Korean companies might be interested in joining our Yemen project. Ray agreed, and in a surprisingly short time that involved one trip to Dallas by the Koreans and three trips to Seoul by our group, an agreement was executed in Seoul between Hunt and Yukong Limited serving as executive agent for a consortium of Korean companies. The consortium consists of four companies including Yukong, Samwhan, Hyundai, and Korea's Petroleum Development Company (Pedco). Terms of the agreement were rather stringent and I shall not forget the tension on 29 February, 1984 as Mr. H. C. So, Vice President of Yukong, signed on behalf of the Korean participants. Once all signatures were in place on all agreements and the agreements were properly documented, I informed Mr. So and Yukong's Executive Vice President, Mr. Hang Duc Kim, that the Alif well had already penetrated into a hydrocarbon-bearing sand that appeared to be of excellent quality, but of as-yet-undetermined thickness. You can well imagine their relief and their sudden transformation from dutiful determination to smiles that quickly grew into outright laughter of elation and gratification. It was one of many happy times relating to Hunt's work in North Yemen.

[I must interject here that we learned of hydrocarbon shows prior to final execution of the Korean participation agreement, and Ray Hunt was reminded that we were not yet bound to the Korean consortium and that it would surely be possible to make an even better deal either with them or with someone else. Ray replied, "No, we've both negotiated in good faith and arrived at mutually acceptable terms. To revert now on the basis of our new information wouldn't be honest!"]

DISCOVERY

The initial well was declared a discovery and assigned the field name of Alif on 20 March, 1984 by Minister of State and Chairman of Yominco, H. E. Abdul Rahman Al-Bahr. Drilling operations continued until 21 June, when at a total depth of 4182 meters (13,720 feet) in basement, plug-back and completion operations began.

Perhaps it was a simple coincidence of timing, or perhaps it was another act of Godly ordination, but testing of the oil zones was undertaken on 4 July, 1984. Regardless of one's choice, the first flow of oil in North Yemen was a strong harbinger of economic independence for the government and people of that country. At the same time, it established Hunt Oil Company as an effective independent international oil and gas contender, if not the current champ!

Potential testing proceeded without problems, yielding a combined oil flow from two zones of 7800 BOPD. Later testing of gas zones recorded 55 million CFGD. A hydrocarbon column of 457 meters (1500 feet) had been defined and the drilling of appraisal and other wildcat wells commenced soon thereafter, proceeding with dispatch and a great deal of enthusiasm.

Since those early days of anticipation and initial success, each subsequent project has been achieved quickly and with a minimum of difficulty. Dry holes have been drilled, of course, but in an insignificant number in light of the numerous discoveries and certainly in terms of the volume of reserves that have been established.

EPILOG

The eight years following discovery at Alif #1 have been demanding, exciting, and fulfilling. Exploration and consequent development have progressed at an accelerated rate such that, at present, more than 15,000 km (9320 miles) of seismic have been acquired. A total of 171 wells have been drilled, resulting in the discovery of eleven fields.

A 10,000 BOPD refinery was designed and constructed by Petro Fac in Tyler, Texas, and then shipped to Yemen for installation at Alif field.

Foster Wheeler, of Houston, Texas, designed a 24"/26" oil pipeline having an initial capacity of 225,000 BOPD. (By adding two more pump stations, this can be increased to 400,000 BOPD.) The pipeline is 423 km (263 miles) long and runs from the Alif field over the mountains to a point near the Red Sea coast village of Ros Isa. Construction was contracted to a consortium of companies that included Consolidated Contractors International, S.A.L. of Athens, Greece; Saipem S.p.A. of Milano, Italy; and Mannesmann Anlagenbau A.G. of Dusseldorf, Germany. Approximately 3000 people were employed in construction of the line and its pump stations. The completed system was dedicated on 9 December, 1987, just $19^{1}/_{2}$ months after construction contracts were signed.

Following the discovery of Alif #1 and the consequent plans and decisions concerning the refinery and pipeline, it was decided to invite

Exxon to participate in the rapidly expanding Yemen operation. On 21 December, 1985, Yemen Hunt Oil Company and Exxon Yemen, Inc., signed a joint venture agreement creating Yemen Exploration and Producing Company. Yemen Hunt Oil Company retained operatorship for all ongoing exploration and production operations.